T0269165

SpringerBriefs in the Mathematics of Materials

Volume 1

More information about this series at http://www.springer.com/series/13533

Susumu Ikeda · Motoko Kotani

A New Direction
in Mathematics
for Materials Science

 Springer

Susumu Ikeda
Advanced Institute for Materials Research
Tohoku University
Sendai, Miyagi
Japan

Motoko Kotani
Advanced Institute for Materials Research
Tohoku University
Sendai, Miyagi
Japan

ISSN 2365-6336   ISSN 2365-6344   (electronic)
SpringerBriefs in the Mathematics of Materials
ISBN 978-4-431-55862-0   ISBN 978-4-431-55864-4   (eBook)
DOI 10.1007/978-4-431-55864-4

Library of Congress Control Number: 2015954958

Springer Tokyo Heidelberg New York Dordrecht London

Printed on acid-free paper

Springer Japan KK is part of Springer Science+Business Media (www.springer.com)

Preface

This book is the first volume of *SpringerBriefs in the Mathematics of Materials* and a comprehensive guide to the interaction of mathematics with materials science. Until recently, the application of mathematics to materials science has progressed mainly to explain *macroscopic phenomena* by means of partial differential equations (PDE). Mathematics however, plays an important role in describing microscopic systems. Recently, direct observation and control of atoms and molecules have become possible and mathematics is expected to describe how macroscopic properties of materials emerge from *microscopic structure*, in particular, from geometrical structure. Furthermore, along with the development of computer technology, a global trend has developed where researchers are deriving important information from a large amount of accumulated data and are designing materials based on this information and mathematics. A new relationship between mathematics and materials science is required.

In the late twentieth century, the importance of injecting mathematics into a wide range of science and technology fields was recognized. Along with a global change in perspective, initiatives bringing mathematics to materials science were encouraged. Materials science has been since its inception an empirical science founded on the many experiments, results, and intuitions. Recently, some researchers have pointed out that it is crucial to *rationalize design and development of materials*, to tighten the development cycle by creating a predictive framework based on theory, and using big data and powerful computational techniques. This book discusses recent attempts to create a *predictive materials science* based on the systematic interaction and cooperation between mathematics and materials science.

In the opening chapters of this book (Chaps. 1 and 2), a selected history of materials science and some examples of the interaction of mathematics with materials science are described. The emergence of materials science was itself a result of the interdisciplinary integration of materials-related disciplines, which occurred in the 1950s and 1960s. Disciplines related to materials include metallurgy, polymer science, ceramics, solid-state physics, and semiconductors. That mathematics is involved in materials science is no wonder. We believe that this historical background will help readers understand the importance of collaborations between mathematics and materials science.

In Chap. 3, we describe some attempts performed at the *Advanced Institute for Materials Research (AIMR), Tohoku University*, to which the authors are affiliated, trying to create a predictive materials science based on a mathematics–materials science collaboration at an institutional level. In this chapter, some examples of attempts and results will be described briefly. To provide researchers a shared, concrete idea of the objectives of mathematics–materials science collaborations, three target projects (1) *Non-equilibrium Materials based on Mathematical Dynamical Systems*, (2) *Topological Functional Materials*, and (3) *Multi-Scale Hierarchical Materials based on Discrete Geometric Analysis* are discussed. These projects were tackled by AIMR researchers and details of their work are presented in subsequent volumes of this *SpringerBriefs in the Mathematics of Materials* series.

In the last chapter (Chap. 4), we describe how breakthroughs based on mathematics–materials science collaborations can emerge. Our argument is supported by the experiences at AIMR where many researchers from various fields gathered and tackled interdisciplinary research.

We understand that a systematic attempt at a mathematics–materials science collaboration needs much time to mature. Such attempts would have been almost impossible if not for the long-term continuous financial support by the *World Premier International Research Center Initiative (WPI)*, jointly managed by the Japanese Ministry of Education, Culture, Sports, Science and Technology (MEXT) and the Japan Society for the Promotion of Science (JSPS), and we sincerely appreciate the support under this program. This attempt also required the enthusiastic effort of our institute. We would like to thank all researchers at AIMR. In particular, we are deeply indebted to Dr. Daniel M. Packwood for his assistance in improving some parts in Chap. 3. Critical reading by Prof. Yasumasa Nishiura and Prof. Masaru Tsukada has greatly improved the manuscript and the authors appreciate their advice. The charts showing the evolution of materials used in Fig.1.3 are mural graphics at the entrance of the new building of the Department of Materials Science and Metallurgy, University of Cambridge. We thank Prof. A. Lindsay Greer and Granta Design Limited in Cambridge for permission to reproduce these graphics. We are deeply grateful to Dr. Paul Tambuyser and Dr. Claude Hootelé for permission to use photographs of goniometers in Fig. 2.2. We are also grateful to Prof. An-Pang Tsai, Prof. Yasumasa Nishiura, and Dr. Akihiko Hirata, Tohoku University, for providing, respectively, a photomicrograph of quasicrystals (Fig. 2.5b), photographs of the BZ reaction patterns (Fig. 2.14), and a schematic explaining the structure analysis of metallic glass (Fig. 3.1). We thank Ms. Miho

Iwabuchi for her wonderful illustration depicting the four steps necessary for achieving interdisciplinary integration used in Fig. 4.4. Finally we wish to thank Mr. Masayuki Nakamura of Springer Japan who helped us with the publication of this book.

We would be very honored if readers of this book were interested in our attempt and join the collaborative network to create new materials science together.

Sendai, Japan   Susumu Ikeda
August 2015   Motoko Kotani

Contents

Chapter 1
A Historical View of Materials Science

Abstract Materials science is an interdisciplinary field of research comprising various areas such as metallurgy, polymer science, ceramics, solid state physics, and semiconductors. The integration of these areas under a single field enables researchers to exchange knowledge, expertise, and experimental techniques and accelerates advances in materials science. As a common language for all fields of science and technology, mathematics can provide a common ground in materials science research and contribute to its development. We begin our journey into materials science with the story of how this field emerged in the 1950s in the United States of America.

Keywords History of materials science · Metallurgy · Polymer science · Ceramics · Solid state physics · Bloch wave function

1.1 Emergence of Materials Science as an Interdisciplinary Field

Rubber, steel, silicon, cloth, glass, ceramics, paper, and wood—these are all materials, each exhibiting characteristic functions and usage. However, although materials science does not investigate the properties of paper and wood, it does investigate the others. For cloth, materials science researches fibers such as nylon, but cloth is seldom studied. This curious circumstance was largely dictated by the historical course taken by materials science (or materials science and engineering, MSE). The rise of "materials science" dates back to the late 1950s when a department at Northwestern University, Illinois, USA, was founded with the name "Department of Materials Science and Engineering" in 1959 [Cah]. On the department's website, one can still read a description beginning "Since it was established as the first materials science academic department in the world more than 50 years ago,...". For Northwestern University, the fields of solid state physics, metallurgy, polymer chemistry, inorganic chemistry, mineralogy, glass and ceramic technology (or sections of them) were integrated under one department thereby creating a new research field made up of existing related disciplines focused on "materials." Because of its historical roots, rubber (polymer science and synthetic chemistry), steel (metallurgy), silicon (solid

© The Author(s) 2015
S. Ikeda and M. Kotani, *A New Direction in Mathematics
for Materials Science*, SpringerBriefs in the Mathematics of Materials,
DOI 10.1007/978-4-431-55864-4_1

state physics and semiconductors) and glass and ceramics (ceramics) became targets for materials science, whereas cloth, paper, and wood were not generally studied in MSE.

From this reorganization, materials science spread around the world. In most universities, disciplines related to materials were integrated within one department renamed "Materials Science" or "Materials Science and Engineering." Of course, the new name reflected individual circumstances and policies of the universities. For example, at the University of Cambridge, the new department was named "Department of Materials Science & Metallurgy," whereas at the University of Oxford it became the "Department of Materials".

Emergence of materials science in Japan

This year (2015) and the next year, Japan will be passing its centennial anniversary of modernization in materials science. We briefly outline its history here. As the first research institute on materials in Japan, the 1st Division for developing nonflammable celluloid and the 2nd Division for iron and steel, was established in 1915 and 1916, respectively, at the Provisional Institute of Physical and Chemical Research of the Tohoku Imperial University. Although the first director of the 2nd Division, **Kotaro Honda** was the inventor of the magnetic KS steel. The institute began with a focus mainly on steel; its founding principle had a broad perspective, "to contribute to the well-being of the human race and the development of civilization through the creation of new materials that are truly useful to society by studying both the application and basic research of a wide range of substances and materials such as metals, semiconductors, ceramics, compounds, organic materials, and composite materials." In 1987, after some organizational change, the division, became the Institute for Materials Research (IMR). Nowadays, many universities have a department of materials science and Japan has some world-renowned national institutes for materials science such as National Institute for Materials Science (NIMS). With a large resource of researchers trained over its 100-year history, Japan has continually pushed research boundaries in both structural materials and electronic materials.

1.2 Classical Fields Within Materials Science

Many materials-related scientific fields predate the 1950s as separate disciplines, such as metallurgy, physical chemistry, chemical physics, polymer science, colloids, ceramics, and solid state physics/chemistry. Around the 1950s and 60s, integration of these separate fields was recognized as providing a multitude of benefits through the large overlap of interests and cross-fertilization of knowledge and expertise. In this way, the new discipline of materials science blossomed into a very comprehensive and interdisciplinary field.

The following describes some of those fields that constitute what can be termed classical materials science. These descriptions will help readers to understand the scope of materials science and its objectives.

Metallurgy

As reflected in the name of the department at the University of Cambridge, Department of Materials Science & Metallurgy, metallurgy held a special position among the fields of classical materials science. It formed the core of materials science in the 19th and 20th centuries. This is understandable because metals have been among the more important materials from prehistoric ages characterized by bronzeware and ironware. Modern investigative techniques and physical considerations were introduced into metallurgy earlier than any other fields [Cot]. For example, metallurgy incorporated the ideas of thermodynamics and phase equilibria (Gibbs' phase rule). Researchers were able to explain formation mechanisms of materials by drawing "phase diagrams" of the systems (e.g., Fig. 1.1).

One of the biggest contributions of metallurgy to society was the invention of the alloy named duralumin by a German metallurgist Alfred Wilm. Aluminum is a beneficial light-weight metal but lacks hardness. In 1903, Wilm discovered that the addition of a small amount (around 4 %) of copper to aluminum remarkably increases its hardness. This new alloy (Al-Cu) was commercialized as "duralumin" in 1909, and it played a central role in the subsequent prosperity of the aviation industry.

Polymer Science

It has been said that the modern lifestyle was largely created after the invention of a single polymer material. Discovered in 1935 by Wallace Hume Carothers at DuPont, U.S.A., nylon (nylon 6,6) is a polymer, a large molecule, which is composed of a number of repeated subunits of an organic molecule. Polymers generally form one-dimensional structures (Fig. 1.2a). Like nylon, polymers can be synthetic fibers exhibiting excellent properties, such as strength, that have never been achieved in natural fibers. Some polymers, such as polyethylene, form two-dimensional or three-dimensional cross-linked structures (Fig. 1.2b) and can be used for making films and containers. As well as metals, polymers are among the most important materials in our modern daily lives.

In the early years of polymer science, chemical synthesis techniques were a main part of research. However, physical concepts were gradually introduced into polymer science making great progress possible. For example, in the 1970s, French physicist Pierre-Gilles de Gennes developed the "scaling concept" of polymers by introducing the equation for exponential growth

$$R(L) = R_0 L^\phi, \tag{1.1}$$

where $R(L)$ expresses the size of the rounded polymer for which the extended length is L, R_0 is a parameter depending on the kind of polymer, and ϕ is the scaling dimension

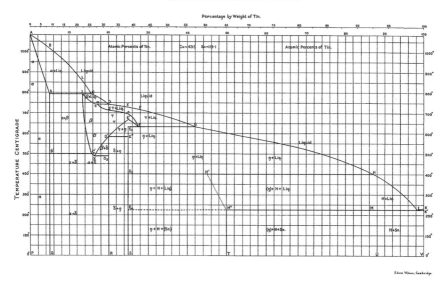

Fig. 1.1 Phase diagram of the Cu-Sn binary system made by Charles Heycock and Francis Neville. Reproduced from [HN] with the permission of the Royal Society

(a) Nylon 6

(b)

Nylon 6,6

Fig. 1.2 **a** Molecular structures of nylon 6 and nylon 6.6 and **b** cross-linked structure of polyethylene

independent of the type of polymer. He applied this concept to the investigation of polymers [deG]. From these beginnings, polymer science has gradually changed from an empirical science to one that is predictive although with some problems difficult to solve. For example, describing the glass transition between the rubbery state and the glassy state still remain unclear.

Ceramics

Ceramics are inorganic materials made through sintering during firing of a raw material mixture. Although ceramics are based on the old technology to make earthenware and porcelain, modern ceramics technology controls the chemical composition of mixtures and temperatures very precisely during firing. Such modern technology can produce fine ceramics which show novel physical, chemical, and electronic properties, and recently ceramics occupy a large part of functional materials [KBU].

Ceramics were fundamentally used in for example tableware, structural materials, electric/thermal insulators, and knives. However, nowadays, many kinds of ceramics have been developed as electronic materials, such as, pressure/gas sensors, actuators, and capacitors. Furthermore, some ceramics show amazing electronic properties. The most familiar is the ceramics composed mainly of copper oxide that exhibit superconductivity at high temperatures. Discovered by Johannes Georg Bednorz and Karl Alexander Müller in 1986, these ceramics have great potential to revolutionize electronic materials as well as materials of daily goods.

Although some readers may dispute the statement, glass and cement are sometimes classified as ceramics. Indeed, ceramics includes a range of materials including earthenware, porcelain, bricks, cement, and glasses, which are collectively termed "old ceramics" to highly refined ceramics, called "new ceramics," "fine ceramics," or "advanced ceramics" that show innovative structural and electronic functionality such as kitchen knives and piezoelectric actuators/sensors.

Solid State Physics and Semiconductors

In the early years of the 20th century, quantum mechanics was developed and refined. One immediate success was describing the dynamics of electrons in atoms, resolving many puzzles of electron behavior. With a fully developed quantum mechanics, the mechanism underlying the bonding of atoms to form molecules and crystals was also explained. In the quantum mechanical treatment of electrons within atoms and molecules, the wave functions for electrons are found by solving the Schrödinger equation. This treatment provided a standard method for the quantum description of electrons in matter. However, condensed matter such as metals and semiconductors are made up of an enormous number of atoms with various kinds of electronic interactions. In practice, solving a many-body Schrödinger equation dealing with all these electrons seemed daunting. For materials science, an extended theory was required to solve the Schrödinger equation describing the behavior of electrons in crystals. For conduction electrons, Bloch's theorem provided a particular solution to this problem. Each crystal has an associated structural lattice (with symmetry belonging to one of the 230 space groups) generating throughout a periodic potential. In 1928, Felix Bloch considered this problem and found the specific solution of the Schrödinger equation compatible with the space group of the crystal lattice. The resulting **Bloch's theorem** can be expressed as follows:

$$\Psi(\mathbf{r} + \mathbf{R}) = e^{i\mathbf{k}\cdot\mathbf{R}}\Psi(\mathbf{r}), \qquad (1.2)$$

where **R** is a lattice vector of the crystal and **k** is a wave vector (wave number corresponding to a momentum) characterizing quantum numbers of the state Ψ. This theorem is a rigorous *mathematical* theorem and strongly restricts the wave functions derived from the Schrödinger equation. The wave functions that satisfy this theorem are called **Bloch wave functions**. To obtain an exact solution of the Schrödinger equation is almost impossible given the complexity of the Hamiltonian describing all energy contributions of the material. Nevertheless, Bloch's theorem enables us to obtain the solution without performing extensive calculations. The Bloch wave functions take the mathematical form

$$\Psi_{\mathbf{k}}(\mathbf{r}) = e^{i\mathbf{k}\cdot\mathbf{r}} u_{\mathbf{k}}(\mathbf{r}) \text{ where } u_{\mathbf{k}}(\mathbf{r} + \mathbf{R}) = u_{\mathbf{k}}(\mathbf{r}). \tag{1.3}$$

Bloch's theorem plays a fundamental role in developing **electronic band theory** by specifying the relationship between energy (ε) and the wave vectors (**k**) of the electrons. This gives the standard approach to understanding the electronic structure and mechanical properties of solids.

Solid state physics is the research field that mainly deals with the electronic structure of crystals described above as well as the atomic/molecular structures, and the electromagnetic properties of solid materials [Kitt]. With the inclusion of certain aspects of metallurgy, it is the largest branch of "condensed matter physics." In the latter half of the 20th century and early 21st century, electronic device technologies based on **semiconductors** have developed to a great extent and have considerably changed daily lives. For this reason, semiconductors occupied a large core of solid state physics research; details are given in Sect. 2.2 and Appendix.

Evolution of Materials

Here we describe an aspect of the advances in materials in connection with the progress of science. The four diagrams in Fig. 1.3 plot the major individual materials used by researchers from the viewpoint of mechanical properties (strength vs density) for the four eras 50,000 BCE, 50 BCE, 1900 (one hundred years ago), and 2013 (present). Naturally, in the era of 50,000 BCE, the materials used were limited to natural substances, such as stone, wood, and some metals that can be obtained without any sophisticated processing techniques. Around the Common Era, the number of types of materials increased because of developments in processing techniques including mining and refining. In the 20th century, the field of materials science became diverse further and subdivided more as it spread.

In the first section of this chapter, we noted the integration of material-related fields into Materials Science that occurred from the late 1950s. Subdividing scientific fields can produce benefits such as deepening the science. However, if subdivision is excessive, one may lose a comprehensive understanding of material characteristics and properties hampering the subsequent development of each subfield. We believe that the connections between subfields are organic in nature and will contribute tremendously in the mutual development of each subfield.

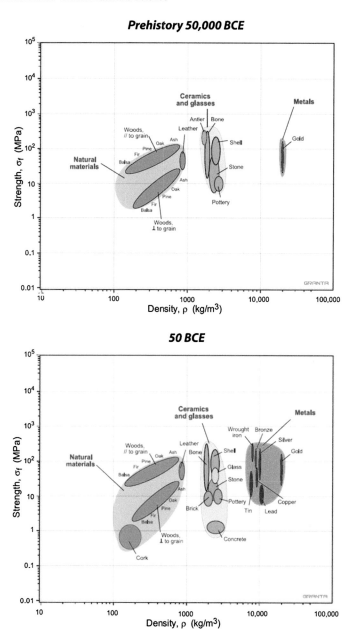

Fig. 1.3 Evolution of materials depicted as density–strength charts. These charts were produced by Granta Design Limited in Cambridge and are currently displayed as mural graphics at the entrance of the new building of the Department of Materials Science and Metallurgy, University of Cambridge. Reproduced by courtesy of Prof. A. Lindsay Greer (University of Cambridge) and Granta Design Limited

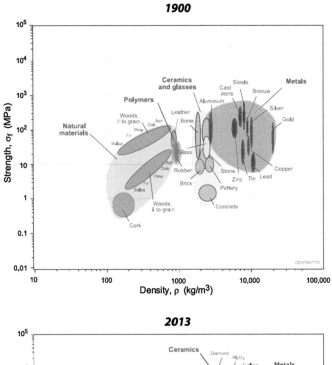

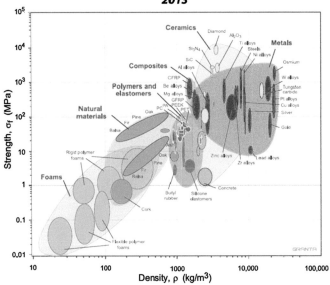

Fig. 1.3 (continued)

1.3 Peculiarity of Materials Science and Partnership with Mathematics

Materials science has an interesting aspect that is slightly different from other disciplines. Generally, most materials scientists are affiliated with departments of materials. In practice, however, materials scientists are to be found in a wide variety of university departments, such as those associated with physics, chemistry, bioscience, geoscience, applied physics, applied chemistry, aeronautical engineering, civil engineering, nuclear engineering, telecommunications engineering, medical science, and pharmacy. This is because every field investigates its own purpose-specific materials. Materials science, more or less, develops strong relationships with all fields of science and technology. In contrast, mathematics is not viewed as partnering well with materials science at present; there appears to be a deep gap between the two disciplines. However, mathematics provides a fundamental, common language for all fields of science and technology. In this sense, mathematics should be a good partner of materials science. In Chap. 2, we present historical evidence that mathematics did contribute to turning points in the advancement of materials science. We wrote this book with the wish that mathematics and materials science develop stronger ties and form a partnership that benefits both.

References

[Cah]   R.W. Cahn, *The Coming of Materials Science*, vol. 5, Pergamon Materials Series (Pergamon, Amsterdam, 2001)

[Cot]   A.H. Cottrell, *An Introduction to Metallurgy*, 2nd edn. (Edward Arnold, London, 1975)

[deG]   P.-G. de Gennes, *Scaling Concepts in Polymer Physics* (Cornell University Press, Ithaka, 1979)

[HN]   C.T. Heycock, F.H. Neville, On the constitution of the copper-tin series of alloys. Philos. Trans. Roy. Soc. Lond. A **202**, 1–70 (1904)

[KBU]   D. Kingery, H.K. Bowen, D.R. Uhlmann, *Introduction to Ceramics*, 2nd edn. (Wiley, New York, 1976)

[Kitt]   C. Kittel, *Introduction to Solid State Physics*, 8th edn. (Wiley, New York, 2004)

Chapter 2
Influence of Mathematics on Materials Science Upto Date

Abstract This chapter discusses the historical influence of mathematics on materials science and mathematical techniques and tools widely used in materials science today. There were many occasions in the past when mathematics and materials science met and interacted to inspire advances. From ancient times, humans are believed to have had an empirical knowledge of materials that was related to mathematics, such as awareness of crystal shape. Such knowledge probably played some basic role in the understanding of materials and in creating new materials. In this chapter, we first focus on the ancients' speculation of *atoms*. The early concept of atoms was produced by some ancient Greek philosophers. Although it took humans more than two thousand years to establish the concept of crystals from the age of the ancient Greeks, the idea of geometrical packing of small building blocks (atoms, molecules, or unit cells in modern science) is supposed to have been somewhere in the humans' consideration. After explaining crystals as one of the most fundamental forms of materials, we will show *quantum materials* and *pattern formation* as the fields with which mathematics has a strong relationship.

Keywords Crystallography · Aperiodic system · Standard realization · Band theory · Quantum Hall effect · Topological insulator · Pattern formation · Crystal growth · Reaction-diffusion equation

2.1 Geometric Structures of Atomic Configurations

2.1.1 Atomism

The first interaction between mathematics and materials is possibly seen in the concept "atomism." In Ancient Greece (5th–4th century BCE), philosophers Leucippus, Democritus, and Epicurus, proposed their two principles of atom and void. They theorized that there exists an ultimate small particle that cannot be divided any further. Naturally, at the time, it was impossible to prove this idea and therefore their atoms remained imaginary.However, it is amazing that they conceived the idea of atoms

© The Author(s) 2015   11
S. Ikeda and M. Kotani, *A New Direction in Mathematics*
for Materials Science, SpringerBriefs in the Mathematics of Materials,
DOI 10.1007/978-4-431-55864-4_2

Fig. 2.1 Five Platonic solids
(polyhedra) drawn by
Johannes Kepler in his work
"Harmonice Mundi (1619)."
Octahedron, tetrahedron,
dodecahedron, cube, and
icosahedron were associated
with air, fire, the universe,
earth, and water, respectively

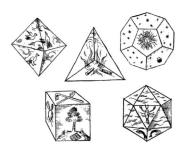

about 2,500 years ago. The ancient Greek philosophers were also mathematicians and therefore it is possible that they thought matter could have a mathematical basis. Democritus likewise said *"All matter is made from invisible atoms like blocks which cannot be separated any more. The universe consists of countless atoms of different shape, size, arrangement and location, and an atom is classified by emptiness."* He believed that real objects can exist because of the existence of emptiness which allows the objects to move without obstruction. Although Plato is not recognized as an "atomist," he developed his own idea of atoms based on geometric objects. He argued that all of creation is made up of five polyhedra (elements): octahedron (air), tetrahedron (fire), dodecahedron (the universe), cube (earth), and icosahedron (water); see Fig. 2.1. He believed all things are made up of an unchanging level of reality based on mathematics. These fundamental polyhedra cannot be divided into smaller parts because they would lose their beauty if they collided with another (e.g., [Cro]).

Unfortunately, this ancient form of atomism was forgotten for more than 2000 years before modern atomism was proposed by John Dalton in the early 1800s. Moreover, more trustworthy evidence for the existence of atoms was presented 100 years later in the theory of Brownian motion proposed by Albert Einstein in 1905 and the diffraction of X-rays by crystal lattices in the early 1910s by Max von Laue, William Henry Bragg and William Lawrence Bragg. W. L. Bragg developed the new technique to calculate atomic arrangements in crystals and this discovery depended largely on his ability in mathematics. Later, with transmission electron microscopes and scanning tunneling microscopes, researchers began at last to observe atoms directly.

2.1.2 The Miracle Year of 1669; The Emergence of Crystallography and Optocrystallography from Mineralogy

In physics, the year 1905 is often called the "miracle year" because Albert Einstein published three important papers, the theory of the photoelectric effect based on the quanta-of-light hypothesis, the theory of Brownian motion, and the special theory of relativity. Here, we describe another "miracle year," which initiated the science of

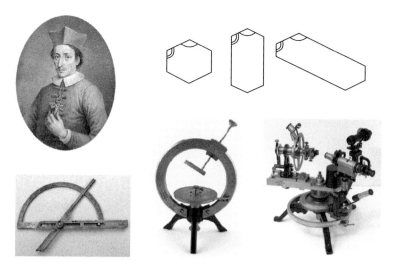

Fig. 2.2 A portrait of Nicolaus Steno (from Wikimedia Commons), the law of constant angle, and development of the goniometer for measuring interplanar angles. Photos of goniometers from the collection at the Virtual Museum of the History of Mineralogy (http://www.mineralogy.eu/index. html) are reproduced here with the museum's kind permission

crystals. Nicolas Steno (1638–86), a Danish scientist of the 17th century, discovered the "law of constant angle" in 1669 (Fig. 2.2). The law stipulates that the interplanar angle formed by two crystal faces (e.g., the angle formed by the (100) plane and (111) plane) are universally constant for the same kind of crystal (mineral). Based on this law, Steno reached the conclusion that crystals grow through the attachment of small particles to crystal facets and the rate of attachment (growth rate) differs for each facet (orientation of the crystal). We see in Steno's discovery definite geometric considerations.

In 1669, another great discovery was achieved by another Danish scientist, Rasmus Bartholin (1625–98). He discovered the "double refraction" of light rays in calcite crystals (Fig. 2.3). This observation formed the basis of the optics of crystals. Before 1669, crystals were mysterious substances and it was beyond human understanding to find mechanisms underlying their fascinating properties. Although descriptive mineralogy existed and the classification of the many kinds of minerals (most minerals are crystals) had been almost completed, the real nature of minerals, for example, the reason why crystals are surrounded by facets, was still unclear. After 1669, however, crystals became a matter for science. We believe that this miracle year, 1669, is the starting point of the modern science of minerals and crystals that subsequently led to materials science. Furthermore, as mentioned above, mathematics, in particular, geometry, played an important role in opening up the new world of materials-related science.

Incidentally, Steno produced another big discovery in 1669—the "law of super-position"—in the same book "*De solido intra solidum naturaliter contento*

Fig. 2.3 A portrait of Rasmus Bartholin (from Wikimedia Commons) and a calcite single crystal showing double refraction

dissertationis prodromus". However, this law is no longer mentioned here because the principle is only remotely related to materials science.

One hundred years after the Steno's discovery of the "law of constant angle", French mineralogist René-Just Haüy (1743–1822) discovered the "law of rational indices." This law states that the indices of any crystal faces are proportional to three small integers. This law also clearly indicates that crystals are composed of small repeating units (building blocks) each with the same shape and same size (Fig. 2.4). Moreover, it provided a microscopic basis for Steno's law showing that the fundamental forms of crystals lie in geometry.

Throughout the 19th century, the understanding of materials greatly progressed. In particular, the field of *crystallography* advanced considerably. For one, given that matter is composed of Daltonian atoms, their atomic arrangements are naturally expressed by *graphs*. A graph is a mathematical structure modeling the interaction (denoted by edges) between objects (vertices). In crystals, atoms are represented by the vertices and interactions by edges. Many materials scientists agree that one of the biggest contributions of mathematics to materials science is Group Theory.

Fig. 2.4 A portrait of René-Just Haüy (from Wikimedia Commons) and his model of a crystal consisting of small building blocks with the same shape and same size. Reproduced from "Traité de Minéralogie (1801)"

The concept of group is a mathematical means to describe the regularities, i.e., point symmetries and periodicities (translation symmetries), inherent in the atomic arrangements (lattices) in crystals. The concept began with Évariste Galois and later extended to the notion of continuous transformation groups by Marius Sophus Lie. Mathematicians played a central role in providing the classification of the 230 space groups belonging to three-dimensional space. This proof was independently established by E.S. Fedorov in 1890, A.M. Schoenflies in 1891, and W. Barlow in 1894. Clearly, mathematics is contributing significantly to the development of materials science through the structural analysis of crystals.

2.1.3 Quasicrystals

In 1982, Daniel Shechtman observed pentagonal symmetry in the electron diffraction pattern (Bragg diffraction) of an aluminum alloy sample using a transmission electron microscope and discovered the existence of quasicrystals [SBGC]. One example of a quasicrystal structure is shown in Fig. 2.5a. All crystals have translational symmetry and hence, by a mathematical principle, pentagonal symmetry is prohibited for crystal structures. Thus, the discovery of pentagonal symmetry surprised many researchers and there were heated discussions on whether quasicrystals were a new type of material or some subspecies of crystals involving twining or unstable intermediate phases. Through the efforts of An-Pang Tsai, many stable quasicrystals were successfully synthesized (e.g., [TIM]; Fig. 2.5b), and quasicrystals attained independent status as a new category of materials. Discovery of these quasicrystals brought about a paradigm shift. Specifically, before their discovery, there were only two classes from the viewpoint of atomic arrangement; one is materials having long-range periodic order and the other is materials having a *random* structure. After the discovery, researchers recognized that long range order does not necessarily mean the

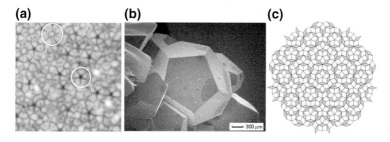

Fig. 2.5 a Scanning tunneling microscopy image (10 nm × 10 nm) showing the fivefold surface of the icosahedral i-Al-Pd-Mn quasicrystal, reproduced from [MSSL] with permission of the Royal Society. **b** A secondary electron image of a Zn-Mg-Dy quasicrystal obtained by scanning electron microscopy, courtesy of Prof. An-Pang Tsai, Tohoku University. **c** One example of a Penrose tiling (from Wikimedia Commons)

existence of translational symmetry, but can involve highly ordered atomic arrangements. Strictly speaking, mathematics did not directly contribute to this discovery. Before the discovery, however, the mathematics community had studied aperiodic tiling. In particular, in 1974, Roger Penrose [Pen] had discovered the so-called "Penrose tiling" (Fig. 2.5c) which is the structure corresponding to the atomic structures of pentagonal quasicrystals. Alan L. Mackay, a crystallographer, who proposed Mackay icosahedra in 1962 [Mac62], predicted in 1981 [Mac81] the possibility of forming quasicrystals associated with the Penrose tiling [Mac82]. More importantly, such tilings suggest where the higher order in quasicrystals originates. De Bruijn [deB] later proved this mathematically by the so-called *cut-projection method*. More specifically, it is an irrational projection of a part (or cut) of the 5-dimensional hyper-cubic lattice projected into a 2-dimensional space. In a sense, mathematics had predicted the possible existence of quasilattices before Shechtman's discovery, but not a physical realization in terms of quasicrystals. This gives an important lesson for both mathematicians and materials scientists. Mathematics has great potential to predict new physical structures of materials and based on these mathematical predictions, materials scientists have good motivation to seek and synthesize some of them. The concept of quasicrystals provides strong support for mathematics–materials science collaboration.

In 1992, a new definition of "crystal" was given by the International Union of Crystallography as "materials whose atomic configuration shows sharp Bragg diffraction peaks and crystals include both ordinary periodic crystals and quasicrystals." [IUC] Hence, quasicrystals fall within the definition from the perspective of point symmetries of the Bragg peaks.

2.1.4 Aperiodic Tiling and Disordered System

A natural mathematical question is to identify the atomic arrangements that admit sharp diffraction peaks. Good references for this subsection are [Sen] and [Ba] as well as references therein.

A mathematical model for the atomic arrangement of disordered materials is a *Delone set*, which is a set of points in a given $\mathbb{R}^n$ satisfying *uniform discreteness* and *relative denseness*. We often consider the corresponding pure point measure (more precisely the autocorrelation measure) of a given Delone set instead of the set itself, and take the Fourier transform of the measure, which provides the mathematical definition of Bragg diffraction for the atomic arrangement. When the Fourier transform has pure discrete measures, the atomic arrangement is called a crystal according to the definition of the International Union of Crystallography. It is an easy consequence of Poisson's summation formula that the atomic arrangement of a classical crystal satisfies the condition.

┌─ X-ray diffraction ───

X-ray diffraction (XRD) is one of the most important analytical techniques in materials science. XRD enables us to determine the crystal structures (arrangement of atoms) and gives useful information about the structure of amorphous materials. In 1912, based on the latest scientific knowledge at that time, Max von Laue conceived the idea that X-rays would be diffracted by crystal lattices because the wavelengths of X-rays are almost the same as the size of unit cells of crystal lattices. Walter Friedrich (research associate) and Paul Knipping (graduate student) proposed an experiment to verify von Laue's intuitive idea and discovered the diffraction of X-rays by crystals. After their discovery, William Lawrence Bragg and his father William Henry Bragg began using the XRD phenomenon to investigate the atomic arrangements in materials and established a fundamental technique to determine the crystal structure of materials. To this day, Bragg's law, $2d \sin \theta = n\lambda$, is still used as one of the most fundamental equations in the analysis of crystal structures. (In the formula, d is the spacing between periodic crystal planes, θ the angle formed by the X-ray beam and the crystal plane, λ the wavelength of the X-rays, and n an integer.) Diffraction takes place only when these parameters satisfy this equation.

Around 1948, about 35 years after his early work, William Lawrence Bragg, as director of Cavendish Laboratory at Cambridge, become interested in the structure of biological substances such as proteins. This period at the Cavendish Laboratory became truly famous when in 1953 James D. Watson and Francis H.C. Crick discovered the double helix structure of DNA. This discovery was based on XRD data and the basic biological fact that the amounts of adenine (A) and thymine (T), as well as the amounts of guanine (G) and cytosine (C), are the same. The story hidden behind this glorious discovery is well-known and presented in many books and documentaries. Although there are various opinions on this matter, it has been mentioned that Maurice H.F. Wilkins (awarded the Nobel Prize in Physiology for Medicine jointly with Watson and Crick in 1962) at King's College showed XRD photographs of DNA that Rosalind E. Franklin had taken to Watson and Crick. Franklin died young in 1958 and therefore was ineligible to receive the Nobel Prize although it was she who had taken the crucial XRD photographs of DNA. Moreover, in 1962, John C. Kendrew and Max F. Perutz of the Cavendish Laboratory were awarded the Nobel Prize in Chemistry for their work using XRD to elucidate the structure of proteins.

XRD is an old but still cutting-edge technology that continually contributes to new discoveries in materials and bio-materials science, even more than 100 years after its development [Sci, Nat].

└──

To this point, we have treated atomic arrangements (point sets) and *tilings* (partitions of the space $\mathbb{R}^n$) interchangeably because there is a geometric correspondence between them. For a given atomic arrangement, we can construct Voronoi cells or dual Delone cells to obtain a tiling, and conversely we can construct an atomic arrangement from a given tiling by taking a representative point in each cell. This correspondence enables theorems from tiling theory to be applied to the study of

materials. The Penrose tiling is an example of an aperiodic tiling. To generalize the
new notion of crystal, we apply the above to a disordered system. We view an atomic
arrangement or a tiling as a dynamical system, as a pair of compact metric spaces
(a completion of the metric space by all translations of the atomic arrangements or
the tilting) and the action of $\mathbb{R}^n$ as translations. The C^*-algebra and its K-theory is a
useful tool to develop quantum mechanics on tilings or their corresponding atomic
arrangements. This approach was initiated by Jean Bellissard, Johannes Kellendonk,
Ian F. Putnam and their followers [Bel86, Kel, KP].

2.1.5   Graph Modeling for Nano-Materials

When one considers interactions between atoms, a *graph theory* is a natural choice
of a mathematical structure to describe their relations. A graph consists of a pair
of sets, one of vertices and another of edges, which represent atoms and bonding,
respectively.

Following the Nobel Prize-recognized discoveries of graphene [Gei, Nov], fullerene
[Osa, KHOCS], and the revolutionary devices based on carbon nanotubes [Iij], the
mathematical studies of possible carbon-networks and their spectral properties have
been most intensive both among the chemistry community and the mathematics community. One attempt involves the spectral study of Schrödinger operators on such
networks. Another involves more general graphs in the context of *quantum graph
models* [BK, EKKST]. Instead of solving partial differential equations (PDE's) for
2-dimensional materials, one can solve and study spectral properties of the corresponding ordinary differential equations (ODE's) on a quantum graph (such as carbon
nanotubes) with proper junction conditions. The mathematical modeling of photonic
crystals has also progressed along many different paths [JJMW, Kuc].

Study of carbon network is one of the more promising research areas. Our discussion of them is continued in Sect. 3.3.

2.1.6   Crystal Lattices and Their Standard Realizations

To study the relationship between microscopic structures in materials and macroscopic properties, *discrete geometric analysis*, i.e., the analysis on graphs, is useful.
Let us introduce an example of the application of discrete geometric analysis on a
crystal lattice. See [KS02, Su13] for details.

A *crystal lattice* is defined as an Abelian cover of a finite graph. Precisely stated, a
crystal lattice is a topological network distinct from its arrangement in physical space
(2-dimensional, 3-dimensional, or n-dimensional). A question posed by Toshikazu
Sunada was whether there is a "standard arrangement" of a given crystal lattice.
Kotani and Sunada [KS00, KS01] have proved that there always exists a unique
standard arrangement, which is called the *standard realization*, and gives the most
symmetric one of a given crystal lattice. For example, the standard realization of
the $\mathbb{Z}^2$-lattices is the square lattice, of the triangular lattice is the regular triangular
lattice, and of the hexagonal lattice is the regular hexagonal lattice.

Sunada further studied and classified all 3-dimensional crystal lattices having the strong isotropic property when the lattices are arranged in their standard realization. Only two structures arise; one is the atomic arrangement of the carbon network of diamond, and the other is the maximal Abelian cover of the K4 graph (K4 lattice) [Su08]. Actually, this structure has been known for almost 100 years to crystallographers and crystal chemists and for over 50 years to materials scientists and solid state physicists, by various names such as the Laves net by Heesch and Laves, Net 1 by A.F. Wells, and **srs** in the Reticular Chemistry Structural Resource, and Y^* in the International Tables for Crystallography (see [HCO] for a history of the discoveries). Nevertheless, from a geometric viewpoint, it is worth mentioning the structure in terms of a "diamond twin," because it has caught the attention of materials scientists. Based on first-principle calculation, Sunada's collaboration with materials scientists predicted the emergence of novel electronic properties for the carbon K4 network [IKNSKA].

Heat flow is a macroscopic feature within materials produced by the microscopic behavior of random walks of particles on an atomic arrangement. Indeed, we derived the heat equation describing its flow over the Euclidean plane by taking a parabolic scaling limit of the random walk on a square lattice. In [KS00], it has been shown that it is possible to generalize this idea to general crystal lattices only when the atomic configuration of the crystal lattice is given by the standard realization. The notion of standard realization has proved to be a natural framework to bridge microscopic structure and macroscopic properties by proposing stable structures of several materials, including Mackay-like carbon networks [TLNKK].

2.2 Quantum Materials

Quantum materials are materials with novel features and properties resulting from the quantum behavior of its electrons. They include, for example, superconductors, magnets, quantum-spin systems, multiferroic materials, as well as conventional semiconductors that have been the mainstay of modern electronics. We do not review the history of Quantum Mechanics throughout the 20th century, but recall that the electronic properties of periodic media (metals, semiconductors, and insulators) are now well understood in principle within "band theory" based on Bloch's theorem. The spectrum and the band structure associated with the wave functions of an electron are obtained by specific Hamiltonians and the solutions of Schrödinger's equation provided by an expansive mathematical knowledge of spectral analysis combined with Fourier calculus.

In real materials, there are many factors that break periodicity, such as impurities, defects, random perturbations because of environmental fluctuations, and the presence of magnetic fields. Hence, in recent studies of quantum materials, much emphasis has been placed on their topological properties. **Topology** is a mathematical tool to aid describing these properties, which display a strong robustness under continuous deformations. It is therefore natural to use K-theory, which studies the topological invariants of vector bundles of manifolds. A good survey for this section is [HK].

2.2.1 Electronic Characteristics of Periodic Materials System: Band Theory

Before discussing the hot topics of quantum materials, we begin by explaining the fundamental relationship between electronic structure and electrical properties of solids. As briefly stated in Chap. 1, Bloch's theorem is the basis from which to consider the behavior of electrons in periodic potentials in crystalline lattices. According to this theorem, the wave function (Eq. (1.3)) of an electron in a crystal is determined by wave vector **k**, which determines the relational expression between energy and wave-number (the magnitude of the wave vector). Because of the scattering or diffraction of electrons resulting from a crystal potential with the translational symmetry of the lattice, Bragg diffraction, for example, produces some energy range forbidden to electrons. The states associated with these energies constitute the "forbidden bands" and the energy width is called the "band gap." In contrast, because there is a large number of valence electrons in solids, the energy ranges excluding the forbidden bands, are filled with allowed energy states and form "allowed bands." In this way, the quantum dynamics of the valence electrons is determined by the combination of allowed and forbidden bands, and the electronic characteristics of solids can therefore be roughly understood using band structure. Using "band theory," the difference in properties of metals, semiconductors, and insulators can be clearly explained, as outlined in Fig. 2.6. In band theory, if the band just below the Fermi level (the valence band) is fully occupied by electrons and if the thermal energy is enough lower than the band gap energy, electrons cannot be excited to the unoccupied bands (conduction band). In this case, the material becomes an insulator, because all electrons in the occupied valence band cannot move. In metals, the Fermi level exists in an allowed band, and the electrons can be easily excited to higher energy states above the Fermi level and become mobile. The probability that an orbital at energy ε is occupied is expressed by the Fermi distribution function (Fermi-Dirac distribution function)

$$f(\varepsilon) = \frac{1}{\exp[(\varepsilon - \varepsilon_F)/k_B T] + 1}, \tag{2.1}$$

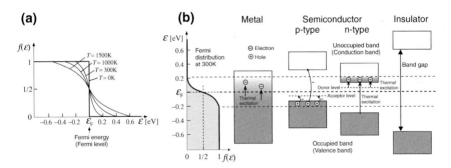

Fig. 2.6 a Temperature dependence of the Fermi distribution function; **b** classification of metals, semiconductors, and insulators based on band theory

where ε_F is the Fermi energy (chemical potential of the system), k_B the Boltzmann constant and T the temperature [Kitt].

For semiconductors, the extent of this function is comparable to the band gap. Then, for n-type semiconductors, electrons in the donor states, as well as a small number of electrons in the valence band, can easily be excited into the conduction band and the material becomes conductive.

For a general introduction to semiconductors and spintronics, see **Appendix**.

Superconductor

Superconductivity was discovered by Heike Kamerlingh Onnes in 1911. In 1908, he was the first to succeed in liquefying helium and thus opened a new research field of low temperature physics. In 1911, with the ability to cool materials, Onnes measured the temperature dependence of electric resistance of mercury. He observed a sudden drop in resistance down to zero when mercury was cooled to 4.2 K. He first thought it was caused by a short circuit, however, he finally understood that the resistance had really disappeared. Thereafter, although the transition temperature of superconductors, T_c, gradually increased, the study of superconductivity did not attract much attention for more than 70 years, despite being of interest to physics, because T_c was actually too low for practical use.

The breakthrough lay hidden in an unsuspecting corner. Johannes Georg Bednorz and Karl Alexander Müller at IBM Zurich Research Laboratory was interested in the relationship between the Jahn–Teller–type lattice distortion and superconductivity. In 1985, when they cooled the La-Ba-Cu-O perovskite, they observed a gradual decrease in resistance at 30 K that almost vanished at 10 K. This material was one kind of ceramics (in general, insulators) and for most researchers it was incredulous that this phenomenon was actually superconductivity. Since some unclear points remained in the experimental data to prove that their new material was a superconductor, Bednorz and Müller subsequently submitted the paper entitled "Possible high T_c superconductivity in the Ba-La-Cu-O system" [BM86, BM87].

Researchers at the Shoji Tanaka Laboratory at The University of Tokyo independently investigated this reported material and finally identified the superconducting phase from the mixture of compounds, thus confirming the first high-temperature superconductor. Historically, a "superconductor fever" followed this discovery. The Nobel Prize in Physics was awarded jointly to Bednorz and Müller in 1987, only 1 year after the publication of their paper.

The mechanism underlying high-temperature superconductivity of the copper oxides (cuprates) is still under debate because its T_c is much higher than that predicted by Bardeen–Cooper–Schrieffer (BCS) theory. Furthermore, a new superconductor series of iron pnictides, discovered by Hideo Hosono's group at Tokyo Institute of Technology [KWHH], has furthered the mystery encountered with superconductivity. We certainly hope that mathematics rises to the challenge and solves this puzzle.

2.2.2   Spin Current

In the past one or two decades, the terminology "spin current" has gradually become one of the important keywords in spintronics. There are two kinds of spin current. One is the spin current accompanied by the flow of charge current; in this case, the current produces Joule heat. The other is spin current without a charge current; in this case, the spin current neither produces Joule heat nor energy dissipation.

We sometimes describe spin as if it has a single rigid direction. However, spin always exhibits a precessional motion (Fig. 2.7a), and this motion can be described by the Landau–Lifshitz–Gilbert–Slonczewski (LLGS) equation (e.g., [BKO])

$$\frac{\partial \mathbf{m}}{\partial t} = -\gamma \mathbf{m} \times \mathbf{H}_{\text{eff}} + \alpha \mathbf{m} \times \frac{\partial \mathbf{m}}{\partial t} + \tau, \tag{2.2}$$

where $\mathbf{m}$ is a unit vector along the magnetization direction, γ the gyromagnetic ratio, $\mathbf{H}_{\text{eff}}$ the effective magnetic field, and τ the current-induced torques. In reality, a huge number of electrons coexist in a system, and one may picture them as many teetotums spinning, each teetotum exhibiting slow swinging (precessional) motion. Such a mass movement of spins actually occurs in ferromagnetic materials and in spin currents. In connection with this spin precession, a new spin current process has been discovered recently. In metals and semiconductors, a spin current is produced based on the migration of each electron having spin. However, in some magnetic insulators such as $Y_3Fe_5O_{12}$, Eiji Saitoh and his colleagues found that the spin-wave mode is a new form of conduction for the spin current [Kaj]. In this case, spin angular momentum can be carried by the collective magnetic moment precession without any migration of electrons (Fig. 2.7b). It is estimated that the spin-wave spin current can carry spin angular momentum up to several centimeters though the relaxation length of spin current is sub-micrometer at most. As well as these spin-wave spin currents, it is believed that a variety of phenomena remains to be discovered like a treasure box with big discoveries yet to be revealed even now.

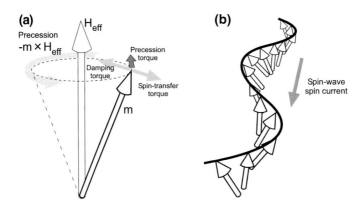

Fig. 2.7   a Precession expressed by the LLGS Eq. (2.2). **b** Spin-wave spin current [Kaj]

2.2.3 Integer Quantum Hall Effect (IQHE)

The IQHE was discovered by Klaus von Klitzing in 1980 (awarded the 1985 Nobel Prize in Physics). Conducted under a strong magnetic field and at low temperatures, it is a phenomenon in which the Hall conductance in a two-dimensional electron system is quantized, exhibiting discrete values $\nu \times \frac{e^2}{h}$ with $\nu \in \mathbb{Z}$, e electric charge and h Planck's constant [KDP].

Expressing the magnetic field B as a closed differential 2-form on a Riemann manifold, there exists a 4-vector potential A, which is a differential one-form satisfying $dA = B$. Then we have a connection $\nabla_A = d + \imath h A$ with which the magnetic Schrödinger operator is written

$$\frac{1}{2m} \nabla_A^* \nabla_A, \qquad (2.3)$$

and acts on a line bundle on the manifold.

In 1982, D. Thouless, M. Kohmoto, M. Nightingale, and M. den Nijs [TKNN] shed further light on the IQHE from a geometrical viewpoint and introduced a topological invariant ν, now called the *TKNN number*, corresponding to the Chern number of the $U(1)$ bundle over the magnetic Brillouin zone.

As a consequence of this topological invariant, special edge states at the interface between two materials with different topological invariants are expected. These states were first identified by Bertrand I. Halperin [Hal] in 1982. Yasuhiro Hatsugai studied the Hall conductance of a system with edges, and defined the *edge index* as the winding number of the edge states on the Fermi surface. He then clarified the bulk–edge correspondence, the relationship between the edge index and the TKNN index (bulk index) [Hat]. The system with random potentials is treated in [BES].

2.2.4 Hofstadter's Butterfly

The tight-binding representation of the 2D electron under a uniform magnetic field acts on $\mathbb{Z}^2$. This situation is usually referred to as the Harper model [Hap];

$$H = U + U^* + V + V^*, \qquad (2.4)$$

where U and V are the so-called *magnetic translation operators*, which satisfy non-commutative relations

$$UV = e^{2i\pi\alpha} VU, \qquad (2.5)$$

for the flux per unit cell of the lattice $\mathbb{Z}^2$.

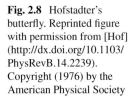
Fig. 2.8 Hofstadter's butterfly. Reprinted figure with permission from [Hof] (http://dx.doi.org/10.1103/PhysRevB.14.2239). Copyright (1976) by the American Physical Society

In 1976, Douglas G. Hofstadter [Hof] revealed its fractal nature as the magnetic field changes. This figure is now known as the Hofstadter butterfly (Fig. 2.8), and attracted much interest in mathematical physics. The *Ten Martini Problems* were posed by Barry Simon, so named after an offer of a martini for each solution by Marc Kac. In 1981, Kac conjectured that the spectrum of the Harper operator is a Cantor set for all irrational α. The problem was solved affirmatively by Artur Avila and Svetlana Jitomirskaya [AJ]. Avila received the Fields Medal for the work.

The Lipschitz continuity of the edge of spectrum in α was proved [Bel94] and generalized for general crystal lattices [Kot03].

Hofstadter's butterfly displays interesting geometric features such as point symmetry and self-similarity. If we consider phenomena encountered in a 2D space subjected to a magnetic field which is parameterized by α, then it becomes difficult to explain. However, if we take the Harper operator as the Laplacian of the 2D non-commutative space where the right/left shift U/U^* and the up/down shift V/V^* are non-commutative, then the geometry of the Hofstadter's butterfly can be understood from an underlying non-commutative space. Non-commutative geometry is thus a natural framework to describe the phenomena.

2.2.5 Central Limit Theorem for Magnetic Transition Operators

The notion of magnetic transition operators was introduced by Toshikazu Sunada [Su94] as a generalized Harper operator. The Harper operator on $\mathbb{Z}^2$ is a discretized magnetic Schrödinger operator on $\mathbb{R}^2$, but the magnetic transition operators are defined on an abstract crystal lattice. Because magnetic fields are differential 2-forms, there is no trivial counterpart on a graph. Sunada used the group cohomology $H^2(\mathbb{Z}^d, \mathbb{R})$ as a substitute of magnetic fields (more precisely a magnetic flux class over the unit cell) and defined a family of magnetic transition operators parameterized by $H^2(\mathbb{Z}^d, \mathbb{R})$. To show the notion is a good discrete analogue, a central limit theorem has been established in [Kot02] forming a bridge between the discrete and continuum. The magnetic Schrödinger operator in $\mathbb{R}^d$ was identified as the scaling limit of the magnetic transition operator.

2.2.6 Topological Insulator

Undeniably, electronics is one of the basic technologies of the modern age. However, instead of using electric currents, we can consider exploiting spin currents by replacing electric charges with spins. The new technology to generate and control spin current is called "spintronics" (details of this topic will be given in Appendix).

In 2004, S. Murakami, N. Nagaosa, and S.C. Zhang predicted spin currents in *spin Hall insulators* [MNZ04]. Because such materials are expected in theory to have no electric current and therefore no energy dissipation, this feature indicates a big potential for applications of spintronics (Fig. 2.9).

In 2005, Charles. L. Kane and Eugene J. Mele [KM] theoretically predicted a *quantum spin Hall effect* for 2D systems with spin up and down electron and time-reversal symmetry (Bernevig, Hughes, and Zhang [BerHZ]); Fu and Kane [FK] in 2006, and independently J. Moore and L. Balents [MB] in 2007, predicted the same for 3D systems. They were confirmed experimentally by L. Molenkamp et al. [Kön] for the 2D case in 2007, and by M. Hasan et al. [Hsi] for the 3D case in 2008. The

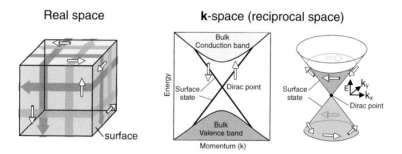

Fig. 2.9 Schematic of a band dispersion for topological insulators

class \ d	0	1	2	3	4	5	6	7	T	C	S
A	$\mathbb{Z}$	0	$\mathbb{Z}$	0	$\mathbb{Z}$	0	$\mathbb{Z}$	0	0	0	0
AIII	0	$\mathbb{Z}$	0	$\mathbb{Z}$	0	$\mathbb{Z}$	0	$\mathbb{Z}$	0	0	1
AI	$\mathbb{Z}$	0	0	0	$2\mathbb{Z}$	0	$\mathbb{Z}_2$	$\mathbb{Z}_2$	+	0	0
BDI	$\mathbb{Z}_2$	$\mathbb{Z}$	0	0	0	$2\mathbb{Z}$	0	$\mathbb{Z}_2$	+	+	1
D	$\mathbb{Z}_2$	$\mathbb{Z}_2$	$\mathbb{Z}$	0	0	0	$2\mathbb{Z}$	0	0	+	0
DIII	0	$\mathbb{Z}_2$	$\mathbb{Z}_2$	$\mathbb{Z}$	0	0	0	$2\mathbb{Z}$	–	+	1
AII	$2z$	0	$\mathbb{Z}_2$	$\mathbb{Z}_2$	$\mathbb{Z}$	0	0	0	–	0	0
CII	0	$2\mathbb{Z}$	0	$\mathbb{Z}_2$	$\mathbb{Z}_2$	$\mathbb{Z}$	0	0	–	–	1
C	0	0	$2\mathbb{Z}$	0	$\mathbb{Z}_2$	$\mathbb{Z}_2$	$\mathbb{Z}$	0	0	–	0
CI	0	0	0	$2\mathbb{Z}$	0	$\mathbb{Z}_2$	$\mathbb{Z}_2$	$\mathbb{Z}$	+	–	1

Fig. 2.10 Periodic table of topological insulators and superconductors. [RT]

materials exhibiting the effect are referred to as *topological insulators (TIs)* because they have conducting states (associated with a spin-polarized Dirac fermion) on the boundary (edge or surface) topologically protected as well as a bulk band gap like ordinary insulators. Other phenomena such as the topological magnetoelectric effect and zero-resistivity in topological superconductors (TSCs) were also found.

The discovery of TI and TSC, and their expression using the more general notion of *topological quantum numbers*, nicely fits in the K-theory framework. A classification of TI and TSC has been given in terms of a symmetry known as the Bott periodicity [AZ, Kita], which gives a one-one correspondence between D-brane charges and the periodic table for topological insulators and superconductors (Fig. 2.10).

2.2.7 Non Commutative Bloch Theory

Let us now consider the quantum mechanics for aperiodic media as an application of K-theory of the C^* algebra. This idea was pioneered by Jean Bellissard [Bel86] using the framework of non-commutative geometry, which was created by Alain Connes [Con]. He received the Fields Medal in 1982 for his work. Non-commutativity has been used to explain the quantization of the quantum Hall conductivities and gap labeling. It is viewed as a promising mathematical framework for quantum mechanics, called *non-commutative Bloch theory* for aperiodic media in general, including periodic media with impurities or defect, quasicrystals, glassy materials, and amorphous materials (cf. [BHZ, KP]).

Using the framework, one derives a bulk–edge correspondence for the IQHE with random potentials [KRS, KSV]. Recently, the notion of *gap index* has been introduced and used to show the bulk–edge correspondence is equivalent to push-forwards via the Gysin map [FHKKMS]. Yasuhiro Kubota has established a general framework for the bulk–edge correspondence of non-periodic Hamiltonians associated with disorder systems using coarse geometry and classified in terms of the K-group of the Roe algebra [Kub].

2.3 Pattern Formation

Generally, materials are created through some chemical reactions or phase transitions. The final products of these chemical reactions/phase transitions are basically determined by thermodynamics, that is, ensembles of materials (phases) in equilibrium appear as final products. However, the textures (e.g., shape of crystals) of the final products show a certain variation depending on the processes involved, including various non-equilibrium and/or non-linear elementary processes (reactions/transitions). Processes forming "patterns" are referred to as *pattern forming* processes, which also influence the properties of materials.

The *calculus of variations* was developed in the 17th–18th century initiated by Pierre-Louis Moreau de Maupertuis. Phenomena in nature are governed by "the law of least action" associated with *the laws of motion and of rest deduced from a metaphysical principle* (1746). The founding fathers of the calculus of variations were the brothers Johann and Jacob Bernoulli, and Johann's student Leonhard Euler. Completion of its foundation was done by Joseph Louis Lagrange. Maupertuis' vaguely formulated philosophy evolved into an elegant mathematical expression, called the *Euler–Lagrange equation* (cf. *Mécanique Analytique* (1788, J.L. Lagrange), and reformulated classical mechanics based on the *variational principles* by William R. Hamilton and Friedrich H. Jacobi.

We need mathematics to describe both spatial shapes in equilibrium and transition dynamics toward equilibrium from non-equilibrium states. We start this section with static pattern formation, which is determined by thermodynamic equilibrium, and gradually precede to dynamic processes such as crystal growth and reaction–diffusion processes. Afterwards, some mathematical ideas to deal with dynamic crystal growth processes, for example, *level set method* and *phase field method* will be described.

2.3.1 Patterns in Equilibrium: Soap Films, Soap Bubbles

Soap films and surfaces of biological or metallurgical cells, are governed by surface tension. Surfaces that model soap films minimize the surface tension energy are in mathematics called *minimal surfaces* with prescribed boundaries that satisfy the so-called *minimal surface equation* derived by Lagrange;

$$\frac{\partial}{\partial x}\left(\frac{f_x}{\sqrt{1 + f_x^2 + f_y^2}}\right) + \frac{\partial}{\partial y}\left(\frac{f_y}{\sqrt{1 + f_x^2 + f_y^2}}\right) = 0, \qquad (2.6)$$

for a graph $z = f(x, y)$.

Historically, the experimental observations by Joseph A.F. Plateau (*Statique Expérimentale et Théorique des Liquides soumis aux Seules Forces Moléculaires* (Experimental and Theoretical Statics of Liquids Subjected Solely to Molecular Forces) published in 1873 inspired mathematics considerably. The Fields medal was awarded to Jim Douglas and Tibor Radó for their solution of the "Plateau's problem" (proving the existence and uniqueness of the minimal surfaces with prescribed boundaries) in 1936.

The self-intersections (singularities) of soap films obey the following

Theorem 2.1 (Intersecting Law) [Tay] *Minimal surfaces intersect over a line at 120 degrees (for three surfaces). Only four such lines can meet at a single point (bringing six surfaces together with the same angles between any two adjacent surfaces).*

It was also conjectured by Plateau and proved by Jean Taylor employing geometric measure theory (developed by Frederick J. Almgren Jr. and many other contributors). The theory gives a complete classification of the local structure of singularities of minimal surfaces. H.A. Schwarz discovered three examples of periodic minimal surfaces; *Schwarz D*, *Schwarz P* and the *Gyroid*. These surfaces appeared as interfaces in phase separations on various occasions in materials science.

Similarly, minimizing surfaces that confine a fixed volume such as soap bubbles, cells, and grain boundaries, cracks of materials, and crevasses in ice are important. They are called *surfaces with constant mean curvature* or *CMC* surfaces in mathematics and give mathematical models of interfaces of two different states: namely *bi-continuous* fields.

Stephen T. Hyde and his collaborators conducted a systematic study of bi-continuous or *poly-continuous* structures of such interfaces in nano-porous materials using minimal/CMC surface theory [HCO, HOP, HO, HR]. Bi-continuous structures have been observed in mesoporous materials, and tri-continuous structures were first synthesized as interwoven porous channels in mesoporous silica by J. Ying et al. [HYZ].

2.3.2 Fundamentals of Crystal Growth

Crystal growth is one of the more important physical and chemical processes in materials science because most of materials is made up of crystals (e.g., [Nis]). Generally, crystals grow from melts, solutions, or vapor phases, and most of the crystals pass through a nucleation process as an initial step in growth. In exceptional situations, some crystals are formed through phase transformations without nucleation such as spinodal decomposition. If we assume that small clusters of radius r are formed in supersaturated vapor or solutions, or supercooled melts, the change in free energy for the formation of the clusters can be expressed as

$$\Delta G(r) = -\frac{4\pi r^3}{3v}\Delta\mu + 4\pi r^2\gamma, \qquad (2.7)$$

where v is the volume of one atom/molecule, $\Delta\mu$ the change in free energy of the system per atom/molecule, and γ the density of the surface energy. The first two terms originate from the volume and surface area of the clusters. If the radius exceeds critical radius r_c, the increase in volume causes a decrease in free energy of the system and enhances the further capturing of atoms/molecules (e.g. [Sun05]). This is the fundamental mechanism of *nucleation* and crystal growth follows after the nucleation. Crystal properties, such as size, shape, amount of impurities, and surface roughness, are largely influenced by the crystal growth process and the process is influenced by a kinetic condition depending on the degree of supersaturation and supercooling as well as a thermodynamic condition. For example, if the supersaturation/supercooling is small, the crystal growth rate is small and the shape of the crystals follows closely the equilibrium shape, which can be determined by the *Wulff construction* (e.g. [Lu]) as shown in Fig. 2.11a. In contrast, if supersaturation/supercooling is large, the growth rate is large and the shape differs from the equilibrium shape. For example, natural pyrite crystals show various shapes depending on the grown environments (Fig. 2.11b). Under larger supersaturation/supercooling conditions, crystals tend to take up more impurities from their environments changing their physical and chemical characteristics considerably.

We shall discuss topics related to crystal growth in several parts of this book because crystal growth is a key issue, which combines microscopic processes of materials and their macroscopic properties and features.

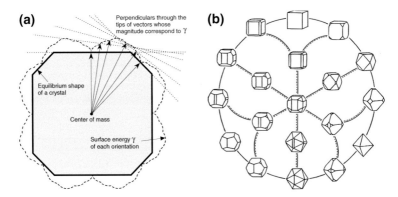

Fig. 2.11 a An example of a Wulff construction. **b** Variations in crystal habit of pyrite [Sun57, Sun05]. Reproduced from [Sun57] with permission of Geological Survey of Japan, AIST

R.L. Dobrushin, R. Kotecký, and S. Shlosman [DKS] proposed in 1989 a microscopic description that uses the Ising model and showed that the Wulff construction appears as a scaling limit. The bridge between microscopic random behaviors of many interacting particles and macroscopic non-equilibrium dynamics of surfaces/interfaces is now studied intensively as the *hydrodynamic limit*. This notion originated with Charles B. Morrey [Mor] (cf. [KL]).

Named after its proposers Mehran Kardar, Giorgio Parisi, and Yi-Cheng Zhang in 1986, the KPZ equation describes the motion of growing interfaces [KZ]. More precisely, this equation is the non-linear stochastic PDE

$$\frac{\partial h}{\partial t}(x,t) = \nu \nabla^2 h(x,t) + \frac{\lambda}{2}(\nabla h)^2 + \eta(x,t), \qquad (2.8)$$

where h is the height of the interface, ν the surface tension, and η white noise (Gaussian), which describes the time evolution of the rough irregular interfaces between vacuum and accumulated material with white noise. The stochastic Burger's equation is a one-dimensional version of the KPZ equation. The combination of non-linearity and randomness causes highly singular behaviors of the solutions and makes a mathematically rigorous treatment of the KPZ equation difficult. Martin Hairer received the Fields medal in 2014 in establishing the theory of regularity structures as applied to the KPZ equation and a more general class of stochastic PDE [Hai].

After describing the diffusion and diffusion–reaction equations below, we shall consider problems of crystal growth again in relation to *mean curvature flow* and some methods describing pattern formation during crystal growth.

┌─ Diffusion ───

Diffusion is a fundamental process in material transport and also important in materials science. From the macroscopic viewpoint, diffusion is a process that homogenizes the concentration; transport proceeds from a high to low region of concentration, and the distribution of concentration gradually becomes uniform. Strictly speaking, the driving force of diffusion is a difference in chemical potential, thus the gradients of temperature and electric potential, as well as concentration, can function as driving forces. Here, however, we consider diffusion as a phenomenon simply driven by a concentration gradient. The most basic formulation of diffusion is given by Fick's law, which in one dimension is expressed in the form

$$\frac{\partial u}{\partial t} = D \frac{\partial^2 u}{\partial x^2}, \tag{2.9}$$

where u is the concentration, t the elapsed time, x the space coordinate, and D the coefficient of diffusion without anisotropy.

Crank published the text book "The Mathematics of Diffusion" [Cra], which assembles and organizes the solutions of the diffusion equation solved under various boundary conditions. For example, if we assume that a thin-film diffusing source of total amount M exists at $t = 0$, the distribution of concentration at time t is given by

$$u = \frac{M}{2(\pi Dt)^{1/2}} \exp\left(-\frac{x^2}{4Dt}\right). \tag{2.10}$$

This distribution is shown in Fig. 2.12a.

In actual systems, for example in nanoporous media, diffusion is not so simple and the concept of "tortuosity" is used (e.g., [Nak]). Tortuosity τ is defined as $\tau = L_p/L$, the ratio of the length of the curve L_p to the distance between its ends L (Fig. 2.12b). As τ increases, material transport is suppressed. In this case, the apparent coefficient of diffusion is approximately expressed using tortuosity τ as

$$D_{app} = \frac{D}{\tau^2}. \tag{2.11}$$

Diffusion occurs in every process involving materials and is an important essential in materials science.

One microscopic aspect of diffusion is the notion of random walks for the atoms. We can observe random walk of atoms by scanning tunneling microscopy and estimate the macroscopic diffusion coefficient (e.g., [MKWL, WCBET, MRPC]). How does the atomic arrangement determine macroscopic diffusions? It is a consequence of the *Central Limit Theorem*. The simple random walk of atoms on the square lattice with parabolic scaling converges to Brownian motion whose distribution law is given by (2.9) as the scaling goes to zero. The classical result was generalized to arbitrary crystal lattices using their standard realization [KS00].

└──

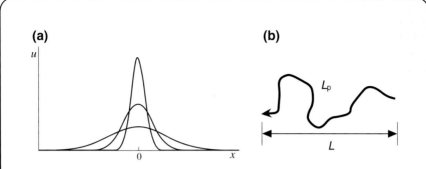

Fig. 2.12 a Time evolution of the concentration distribution based on the Eq. (2.10); **b** schematic of material transport through a complicated path

2.3.3 Reaction–Diffusion Equation

Turing Pattern

Alan Turing opened new insights into pattern formation in nature. In 1952, he claimed in his paper titled "The chemical basis of morphogenesis" [Tur] that pattern formation in biological systems is controlled by chemical reactions. The notion of *diffusion-driven instability*, namely, "non-uniformity may arise naturally out of a homogeneous, uniform state," introduced by Turing, was surprising because the diffusion usually brings a homogeneous stable state. A certain system consisting of two chemical components with different coefficients of diffusion causes instability. Non-trivial patterns can arise in a homogeneous system as a result of competition between reaction and diffusion.

A basic set of the reaction–diffusion equations for such pattern formation is expressed as follows:

$$\begin{aligned}
\frac{\partial u}{\partial t} &= f(u, v) + D_u \nabla^2 u \\
\frac{\partial v}{\partial t} &= g(u, v) + D_v \nabla^2 v,
\end{aligned} \tag{2.12}$$

where $u = u(r, t)$ and $v = v(r, t)$ are concentrations of the two chemical components, D_u and D_v are the coefficients of diffusion of the two components, and f and g are chemical reaction rates.

If the coefficients of diffusion for two components are quite different and the reaction terms f and g satisfy some condition, these equations spontaneously produce characteristic patterns seen in nature, including patterns of some animals and fishes

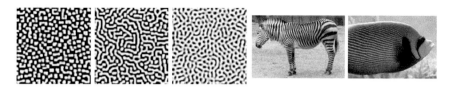

Fig. 2.13 Simulated Turing patterns and patterns seen on a zebra and an emperor angelfish. The first three simulated patterns are reproduced from Wikimedia Commons and the latter two photographs are from Pixabay

(Fig. 2.13). These patterns, which are known as *Turing Patterns*, originate from some instability depending on the delicate balance between reaction and diffusion.

BZ Reaction

A biochemist Boris P. Belousov in 1958 [Bel] proposed a system with more than two components to explain an oscillating chemical reaction. This was confirmed experimentally by Anatol M. Zhabotinsky [Zha] and the equation became known as the *BZ (Belousov–Zhabotinsky) equation*. Rich patterns are generated under the same physical environment but slightly different initial conditions (Fig. 2.14).

A deep question is how patterns are formulated and how they are selected from several stable solutions of the equations. Little is known in general. It may be answered by looking at dynamical properties, hidden structural stabilities, and global attractors. A good reference for the following discussion is [N].

There are four classes of reaction–diffusion systems (Eq. (2.12)):

- Turing system,
- Oscillatory system,

Fig. 2.14 Formation of an oscillatory spiral pattern from the BZ reaction. Courtesy of Dr. Tatsunari Sakurai (Chiba University) and Prof. Yasumasa Nishiura (Tohoku University)

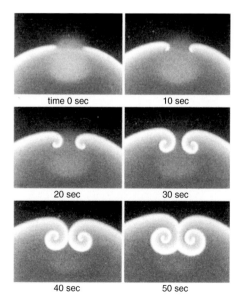

- Excitable system,
- Bi-stable system,

depending on the dynamical nature of the corresponding kinetic equations $u_t = f(u, v)$ and $v_t = g(u, v)$.

The bi-stable system has two attractors whereas the others have one. The oscillatory system has a limit cycle as its attractor. As we have already seen, the Turing system has a diffusion-driven instability; this occurs because the stable point of the kinetic equation is destabilized by the diffusion terms. The excitable systems exhibit the more interesting phenomenon such as the BZ reaction, Rayleigh–Bénard convection, and nematic liquid crystals.

Cahn–Hilliard Equation

John W. Cahn and John E. Hilliard in 1958 [CH] proposed the following equation (Cahn–Hilliard equation), which is one of the basic reaction–diffusion equations and describes the process of phase separation such as spinodal decomposition:

$$\frac{\partial u}{\partial t} = D\nabla^2 (u^3 - u - \gamma \nabla^2 u) \tag{2.13}$$

where u is the concentration, D the coefficient of diffusion, and γ a parameter indicating the length of the transition regions between two kinds of domains. The two separated phases are expressed as $u = 1$ and $u = -1$. In a spinodal decomposition, small fluctuations in the concentrations of mixed components gradually spread and domains of the two phases develop without nucleation (Fig. 2.15). There are variations of the *Allen–Cahn equation* [AC] and their stochastic versions.

Far-From-Equilibrium Dynamics

Ilya Prigogine (Chemistry Nobel laureate in 1977) expanded Turing's idea into a new broader field of science, called "dissipative structures." A dissipative structure is a self-organizing structure, which appears in a dissipative system with steady-state going-in and out of energy. Such a state is achieved under far-from-equilibrium conditions [NP]. A typical example is any living organism such as our bodies. Our bodies are open systems, which take energy (foods) in and discharges heat and excretion. Inside bodies, there are some chemical reactions, which seem to oppose

Fig. 2.15 Evolution of phase separation simulated based on the Cahn–Hilliard equation. Captured snapshots from Wikimedia Commons

thermodynamics. Nevertheless, our bodies regulate continually toward a steady state. Prigogine and his colleagues thought that these systems are among the patterns that appear under far-from-equilibrium conditions.

The relationship between structure and function and that between structure and process are key issues in materials science. Revolutionary functions often emerge from intricate structures, and those structures are created in manufacturing processes under far-from-equilibrium conditions such as rapid cooling. Examples can be seen in condensed matters like glasses, powders, foams, granular materials, dense colloidal materials, polymers, and various particular systems of different sizes. One can say that the invention of processing techniques for novel materials resulted from explorations of how we can control far-from-equilibrium dynamics and pattern formation. Far-from-equilibrium dynamics is ubiquitous [N]; it is not only found in materials science and condensed matter physics but also surfaces in challenging issues encountered in biology, chemistry, geoscience, meteorology, astronomy, and social sciences. The mathematical study of far-equilibrium dynamics has just began.

2.3.4 Mean Curvature Flow to Describe Crystal Growth

Mean curvature flow simply describes the movement of a curved plane for which the velocity along the normal direction is proportional to the mean curvature around each point on the surface. The concept of mean curvature flow was originally proposed in a model for the formation of grain boundaries in the annealing of pure metal by materials scientist William W. Mullins [Mul], and 20 years later Kenneth A. Brakke established a mathematically rigorous framework [Bra] (cf. [GG]).

Mathematically, the equation for mean curvature flow is difficult to solve because singularities appear in a finite time given smooth initial values. The **level set method** has been developed to overcome this difficulty. The idea first appeared in [OJK] and later introduced by Osher–Sethian in numerical calculations [OS].

Another approach to treat mean curvature flow is to use the "viscosity solution," a generalized notion of classical solution of the PDE developed by Pierre-Louis Lions and Michael G. Crandall. In [CGG], the viscosity solution was proved to be useful in analyzing the mean curvature flow.

When we consider surfaces or interfaces with widths, the **phase field method** is efficient. The method introduces an auxiliary field that takes values 1 inside, gradually changing to -1 outside. Taking the scaling limit of the phase field equation, we obtain equations of the sharp surfaces and interfaces appearing in the Stefan model and Mullins–Sekerka model. The method provides models for needle-like crystals or dendritic crystals with side branches [Kob93, Kob94].

The following gives a brief introduction to the **level set method** and **phase field method**.

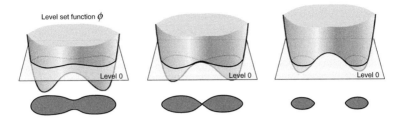

Fig. 2.16 Schematic explaining the concept of level set method

2.3.5 Level Set Method

Stanley J. Osher (2014 Gauss Prize winner) and James A. Sethian developed "*level set method*" to compute moving fronts involving topological changes. In the level set method, the curved surface is regarded as a contour at level zero of a certain function ϕ (Fig. 2.16), and the equation of motion of the contour is replaced by a PDE of ϕ. This PDE can be solved using the viscosity solution method.

One of the successful examples of its application to materials science is the simulation of epitaxial growth of thin films. A simulation of crystal growth is often carried out as an atomistic process, for example, using the molecular dynamics (MD) method or Monte Carlo method. In such cases, the computation load is very high because it has to treat thousands of atoms individually. The level set method simulates only the boundary between the underlayer and overlayer and does not treat each atom. This simulation seems rough compared with other simulation methods using atomistic process. However, the method is based on rigorous mathematics and yields some benefits. Also, in comparison with results from other methods, the level set method has been proved to derive accurate results, for example, in the distribution of island sizes in epitaxial growth [CR].

2.3.6 Phase Field Method

If a solid/liquid interface moves during solidification (cooling), the following three main laws apply: (1) curvature effect (Gibbs–Thomson effect), (2) conservation law of substances and energy (Stefan condition), and (3) diffusion of heat and substances in solid and liquid driven by a temperature or concentration gradient. However, it is very difficult to solve its time evolution not only analytically but also numerically. The phase field method was developed to solve such a problem.

The basic phase field model describing solidification and melting is

$$\frac{\partial u}{\partial t} + \lambda \frac{\partial \phi}{\partial t} = \nabla^2 u$$

$$\alpha \varepsilon^2 \frac{\partial \phi}{\partial t} = \varepsilon^2 \nabla^2 \phi + f(\phi, u; \varepsilon),$$

(2.14)

where ϕ is the order parameter indicating solid or liquid ($\phi = -1$ and 1 indicate solid and liquid, respectively). The characteristic difference from other models (e.g., Stefan's model) is that the phase field model deals with a solid/liquid interface having some thickness (a "diffuse interface" is one for which ϕ ranges from -1 to 1) whereas Stefan's model deals with the interface as a discontinuous sharp boundary. This means that the phase field method does not treat the interface precisely (i.e., the precise position of the interface, which changes from one moment to the next, does not need to be recorded). Hence, this method makes it possible to generate complicated structures such as dendrites.

To simulate the growth of dendritic crystals, including the effect of anisotropy, Ryo Kobayashi [Kob94] used the following equations

$$\frac{\partial u}{\partial t} = \nabla^2 u + \frac{\partial \phi}{\partial t}$$

$$\tau \frac{\partial \phi}{\partial t} = \varepsilon^2 \nabla^2 \phi + \phi(1-\phi)\left(\phi - \frac{1}{2} + m(T, -\nabla\phi)\right),$$

(2.15)

$$m(T, v) = -c \tan^{-1}(\gamma \sigma(v)T)$$

$$\sigma(\theta) = 1 - \frac{1}{4}\delta(1 - \cos 4\theta)$$

(2.16)

where δ is a parameter controlling the degree of anisotropy. In changing its value from 0 to 1, the pattern for the grown crystal changes from coral-like to needle-like via dendritic shapes (Fig. 2.17a). This morphological change can be seen in various materials systems. For example, organic semiconductor pentacene shows dendritic structure (Fig. 2.17b) if a thin film of pentacene is grown on an inert flat substrate (e.g., SiO$_2$) by molecular beam deposition method in vacuum.

These mathematical formulations for pattern formation have played an important role not only in understanding the pattern formation process but also in fostering ideas in controlling non-equilibrium and/or non-linear phenomena and designing structures of materials.

(a)

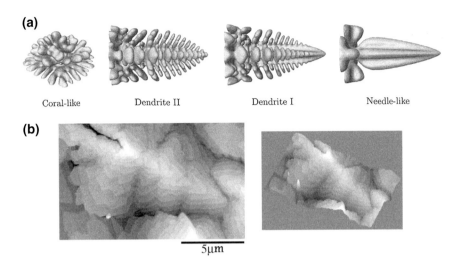

    Coral-like   Dendrite II   Dendrite I   Needle-like

(b)

5μm

Fig. 2.17 **a** δ-dependence of the morphological variation using the phase field model. Reproduced from [Kob94] with permission from Taylor & Francis Group. **b** Topographic image obtained in atomic force microscopy of a thin film of the organic semiconductor *pentacene* grown on a silicon substrate with oxidized surface. The molecular steps for which the height almost corresponds to the length of a pentacene molecule (about 1.5 nm) are clearly seen. The 3D image was constructed by ImageJ software (U.S. National Institutes of Health; http://imagej.nih.gov/ij/). Please see [YTNK, PBNZ, SS] for the morphological variation of pentacene thin films

2.4 Other Tools

In this section, we introduce certain technologies, which are important in materials science, where mathematics plays a critical role in their realization.

2.4.1 Computed Tomography

Imaging techniques are crucial in the development of materials science. Nowadays, through their advancement, as with three-dimensional (3D) imaging, many important results have been accumulated regarding structure and formation of textures. X-ray computed tomography (CT) is one powerful technique used to observe 3D internal structures of materials. The fundamental principle in CT was given by Austrian mathematician Johann Radon in 1917. He invented the Radon transform which enables the reconstruction of 3D data from an infinite set of projections. The intensity of the transmitted X-ray through the object is

$$I = I_0 \exp\left(-\int \mu(x, y)ds\right), \tag{2.17}$$

Fig. 2.18 Absorption of a
parallel X-ray beam by an
object

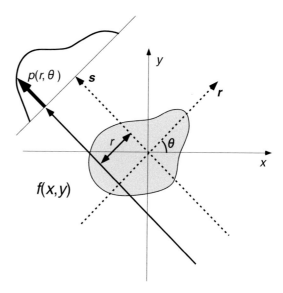

where I_0 is the intensity of the incident beam and $\mu(x, y)$ is the coefficient of attenuation for each point in the object (Fig. 2.18).

The intensity of the projection $p(r, \theta)$ can be expressed using the Radon transform,

$$p(r, \theta) = -\int_{-\infty}^{\infty} \mu(r \cos \theta - s \sin \theta, r \sin \theta + s \cos \theta) ds \qquad (2.18)$$

and coefficient of attenuation for each point $\mu(x, y)$ can be obtained by the inverse Radon transform.

A computation technique for X-ray CT was first proposed by Allan M. Cormack in 1963 and 1964 [Cor63, Cor64]. He established the tomographic calculation procedure, which repeats the calculation to obtain discrete data using the Radon transform and its inverse. In 1971 (the paper was published in 1973), Godfrey N. Hounsfield succeeded in developing a practical X-ray CT scanner and obtaining tomographic images of the human brain based on Cormack's theory [Hou]. That few researchers were aware of Cormack's theory before Hounsfield's great invention of the century is a famous story.

An example of an X-ray CT image is shown in Fig. 2.19. X-ray CT involves non-destructive measurements enabling internal 3D structures to be obtained without damaging valuable samples. Such measurements are very important in medical diagnoses. The X-ray CT scanner exemplifies the successful industrial invention that is derived from the fruits of modern mathematics.

The field of X-ray CT is continuously making remarkable progress in the development of new mathematical algorithms. For example, for sparse image reconstruction from a limited number of projections and interior CT that can produce 3D images of

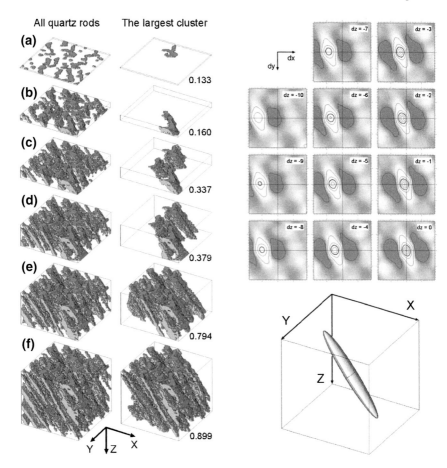

Fig. 2.19 *Left*: An example of X-ray CT data showing the 3D connection of quartz crystals in natural graphic granite, and *Right*: Expression of the average 3D shape of quartz crystals using two-point correlation function $AC(\delta) = \langle B(\mathbf{r} \cdot B(\mathbf{r} + \delta)\rangle_{\mathbf{r}}$. Republished with permission of the Mineralogical Society of Great Britain and Ireland, from [INN]; permission conveyed through Copyright Clearance Center, Inc

a region of interest (ROI) using limited projections of the ROI (e.g., [KCND, KSR]) as well as in the development of hardware.

2.4.2 Some Other Computational Tools

- **Fast Fourier transform (FFT)**
  In spectral measurement in materials science, in particular, infrared spectral measurement, the FFT is indispensable enabling a frequency-absorption intensity

relationship to be obtained very quickly using white light without any mono-chromatized light and a spectroscopic mechanical system. The algorithm for the FFT was developed by American mathematicians, James W. Cooley and John W. Tukey, in 1965 [CT]. Strictly speaking, historical investigations suggest that Carl Friedrick Gauss had invented a similar algorithm around 1805. Nevertheless, they succeeded in decreasing the computational load dramatically by using the symmetry associated with the discrete Fourier transform (DFT).

Like X-ray CT, Fourier transform infrared (FTIR) spectroscopy is another very successful industrial invention based on the fruits of modern mathematics.

- **Nonlinear pattern fitting (peak separation)**

  In spectral analysis, we often face the overlapping of some peaks, for example, in X-ray diffraction measurement, and need to separate them into their individual components. In such situations, we apply nonlinear pattern fitting to separate peaks. In the computational procedure, we first have to decide the appropriate function that expresses the shape of the single peak. For example, the following normalized split (asymmetrical) pseudo-Voigt function, a combination of Gaussian and Lorentzian functions, is frequently used to fit to the shape of each peak (e.g., [Tor]):

$$y(x) = I \frac{(1+A)[\eta_{high}+(1-\eta_{high})(\pi \ln 2)^{1/2}]}{\eta_{low}+(1-\eta_{low})(\pi \ln 2)^{1/2}+A[\eta_{high}+(1-\eta_{high})(\pi \ln 2)^{1/2}]}$$
$$\times \left\{ \eta_{low} \frac{2}{\pi W} \left[1 + \left(\frac{1+A}{A}\right)^2 \left(\frac{x-T}{W}\right)^2 \right]^{-1} + (1-\eta_{low}) \frac{2}{W} \left(\frac{\ln 2}{\pi}\right)^{1/2} \exp\left[-\left(\frac{1+A}{A}\right)^2 \ln 2 \left(\frac{x-T}{W}\right)^2 \right] \right\}, (x \leq T),$$

$$(2.19)$$

where I is the integrated intensity of the peak, T the peak position, W the full-width at half-maximum (FWHM) of a peak, η_{low} and η_{high} denote the ratio of the Lorentzian component at the lower and higher side of the peak, respectively, and A is another asymmetry parameter. Pattern fitting is essentially the process mini-mizing the sum of the squared residuals and the difference between the observed intensity at each position $y(n)_{obs}$ and the calculated value $y(n)_{calc}$ using the above function. First we calculate the partial differential coefficients with respect to I, T, W, η_{high}, η_{low}, and A for each position. (When the number of peaks is more than one, the treatment of the coefficients becomes slightly complicated.) All the partial differential coefficients become elements of a matrix A and the optimum values for parameters I, T, W, ... can be directly obtained by calculation with matrix equation

$$x = (A^t W A)^{-1} A^t W_y, \tag{2.20}$$

where W denotes a weighting matrix.

For non-linear fitting, an iterative calculation (e.g., Gauss–Newton method) is necessary to find the best solution. Examples of X-ray pattern fitting [IIK94, IIK95] are shown in Fig. 2.20. Stable convergence is often difficult to achieve in iterative calculations and various mathematical algorithms, for example, the modified Marquardt method [Mar], have been proposed to guarantee a stable and rapid convergence. Recently, a novel method of spectral deconvolution based on

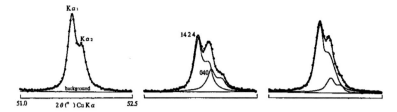

Fig. 2.20 Some results of least-squares fitting of X-ray diffraction patterns reproduced from [IIK94]

Bayesian estimation with the exchange Monte Carlo method has been proposed to avoid a solution trapped in a local minimum because of system hierarchy and nonlinearity [NSO].

In Ancient Greece, there was no strict boundary between mathematics and other fields. Philosophers were mathematicians and meditated on anything from a deep mathematical perspective. Also in the early modern period, scientists, such as Isaac Newton, had all-around abilities and achieved great discoveries not only in mathematics but also in physics and chemistry. This implies mathematics is always central in all considerations. It is easy to imagine that mathematics or a mathematical perspective also influences materials-related sciences. A few centuries later, science has evolved into specialist fields. The wall between individual fields, especially between mathematics and other fields, is becoming wider and higher. Unlike Greek philosophers, there is no polymath with a broad knowledge of various fields. Of course, physicists use mathematics to describe theories and solve equations. However, in modern times, even physics seems to be developing independently of mathematics. In the past one or two decades, research institutions, departments, and governmental programs with aims to encourage interdisciplinary integration and create new scientific fields have been established. These attempts are based on the expectation that such meetings between different fields of knowledge will inspire new ideas not readily produced in a single-field environment with its limited knowledge and experiences. Even in physics, in which mathematics is in use every day, interactions with mathematicians will surely produce new ideas and findings.

In the following section, we shall discuss a recent trend in mathematics–materials science cooperation and collaboration. In just a few decades, it has been recognized that rationalizing design and development of materials is important as is the tightening of the development cycle by the creation of a predictive framework based on theory and mathematics. Various attempts have been proposed around the world to encourage the interaction of mathematics with other disciplines, as well as industry, to encourage innovation. A mathematics–materials science collaboration is one of those attempts.

2.5   Global Trend to Encourage Mathematics–Materials Science Cooperation

In the late 1990s, a movement emerged to promote interactions between mathematics and other fields of science and technology as well as industry. In 1998, the National Science Foundation (NSF) published what is often called the *Odom Report*. Its full title was "*Report of the Senior Assessment Panel for the International Assessment of the U.S. Mathematical Sciences*" and William E. Odom, Lieutenant General, USA (retired) was the chairman of the committee making the assessment. In this report, mathematical sciences were described as being traditionally active in Europe, but many talented mathematicians born in Europe were working in the US. As a consequence, the US had the strongest national community in the mathematical sciences in the world. This implied that US mathematics is strong but somewhat fragile. This report also noted that financial support going to mathematical sciences was lacking and there was some delay in promoting collaborations between mathematicians and users of mathematics similar to the leading-edge approach undertaken at the Isaac Newton Institute in the University of Cambridge. After the publication of this report, the NSF began actively encouraging interactions between mathematics and other disciplines with financial support. One typical example of the new direction can be seen in the Institute for Mathematics and its Applications (IMA) at the University of Minnesota. The IMA was established in 1982 with the objective of promoting synergies between mathematics and its applications. The NSF rated the IMA's activities highly and increased its budget over the period 2005–2010 to accelerate progress. Another example is the Institute for Pure and Applied Mathematics (IPAM) established at University of California, Los Angeles (UCLA). The IPAM was founded in 2000, just after the publication of the Odom Report. Its mandate is to foster the interaction of mathematics with all of science and technology, to build new interdisciplinary research communities, to promote mathematical innovation, and to engage and transform the world through mathematics. It offers two or three 3-month-long programs every year as well as shorter programs such as 1-week workshops. Both the IMA and IPAM are among eight NSF Mathematical Sciences Research Institutes (http://www.mathinstitutes.org/). As well as the Isaac Newton Institute, IMA, and IPAM, the Lorentz Center at Leiden University, the Netherlands, provides a good environment for international workshops at the forefront of the sciences including mathematics.

   With respect to applied mathematics, in 2008 the US Department of Energy (DOE) published "*Applied Mathematics at the U.S. Department of Energy: Past, Present and View to the Future*", which has come to be known simply as the *Brown Report*. In regard to energy policy, applied mathematics is more important. In particular, computational simulations of large-scale complex systems, such as regional and global climate, fluid dynamics of water, and electromagnetic fields of plasma (for nuclear fusion), are the main targets. Risk analysis is also significant. They argued that mathematics is important in improving models and overcoming the difficulties solving problems, such as the uniqueness and reconstruction methods for inverse problems. The *Multiscale Mathematics Initiative* is another program that the US DOE promoted in 2000. Macroscopic phenomena and properties of matter arise from the microscopic processes of atoms and molecules. However, there are many hierarchical layers between the micro- and macro-scale, and so far, different theories and models used in computations have been developed to analyze the various phenomena that exist in each hierarchical layer. Multiscale mathematics aims to create a unified framework for the simulation of macroscopic phenomena by combining all physical and chemical processes from the microscale to the macroscale. The roadmap to create multiscale mathematics was compiled in 2004 and the program was carried out over 3 years, from 2004 to 2006. About 60 % of the budget was approved for topics from materials science and mathematical methodologies and tools applicable to, for example, the creation of nanomaterials. Micro/nano-fabrication techniques for integrated circuits were developed.

In 2008, the OECD published *"Report on Mathematics in Industry."* In its introduction, the report clearly noted that industrial innovation is increasingly based on results and techniques of scientific research, and that much research is underpinned and driven by mathematics. They organized the *Global Science Forum (GSF)* and discussed mechanisms for strengthening the connection between mathematics and industry. They presented results of their assessment at the International Congress on Industrial and Applied Mathematics (ICIAM) conference (2007). They also discussed these problems with mathematicians and applied mathematicians. Through these attempts, the importance of the interaction between mathematics and industry is gradually making in-roads among policy makers, (applied) mathematicians, and industry people. Indeed, some big projects promoting the application of mathematics such as *Mathematics of Planet Earth 2013* and the Deutsche Forschungsgemeinschaft (DFG) Research Center Matheon, are progressing. Some books reporting on the present status of the mathematics–industry cooperation have been published [LER, GK].

In recent years, the importance of the use of *"Big Data"* is increasing year by year. Also, in materials science, it has been pointed out that it is crucial to rationalize the design and development of materials and tighten the development cycle by constructing databases of research results related to materials, and analyzing Big Data from Informatics Theory and Mathematics. In 2011, the US launched the *"Materials Genomics Initiative" (MGI)* to develop infrastructure to accelerate the discovery of advanced materials, especially through the use of computational capabilities, data management, and an integrated approach. Described in the following sections, a pioneering approach by the Advanced Institute for Materials Research (AIMR), Tohoku University, to building a mathematical foundation to predict structure and function of materials is quite timely in putting forth highly-functional numerical calculations for predictive capabilities of materials properties.

Another US program, *"Science at the triple point between mathematics, mechanics and materials science,"* started in 2011, is up and running. The project run by the Partnerships for International Research and Education (PIRE) is a 5-year NSF-funded international collaboration program that began in 2011. The main topics come from applied mathematics and mechanics, including partial differential equations, calculus of variations, and scientific computation, which find use in materials science.

In Japan, a movement toward collaborations between mathematics and other fields was slightly delayed. This was noted in the report *"Mathematics as deserted science in Japanese S&T policy,"* published by the *Ministry of Education, Culture, Sports, Science and Technology (MEXT)*, Japan. In recent years, however, this movement quickly gained momentum through the establishment of the *JST CREST/PRESTO* program *"Alliance for Breakthrough between Mathematics and Sciences (ABMS)"* led by Yasumasa Nishiura. In this program, various research proposals related to materials science have been accepted and new fields for mathematics–materials science collaboration are beginning to open up. The accepted proposals include:

- CREST

    - A Mathematical Challenge to a New Phase of Material Science (Team Leader: Motoko Kotani)

- PRESTO

    - Resolution of Fine Structures of Block Copolymers by Young Measure and its Development (Yoshihito Oshita)
    - Global Analysis on Geometric Variational Problems and its Applications (Miyuki Koiso)
    - Development of New Dynamical Simulation Methods by Applying Mathematics (Naoshi Ichinomiya)
    - Mathematical Physics on Interfacial Tension in Nonequilibrium Systems (Hiroyuki Kitahata)

Reference Websites

- "Report of the Senior Assessment Panel for the International Assessment of the U.S. Mathematical Sciences" (Odom Report), National Science Foundation (NSF)
  http://www.nsf.gov/publications/pub_summ.jsp?ods_key=nsf9895
- "Applied Mathematics at the U.S. Department of Energy: Past, Present and View to the Future," (Brown Report), U.S. Department of Energy (DOE)
  http://science.energy.gov/~/media/ascr/pdf/program-documents/docs/brown_report_may_08.pdf
- OECD Report "Report on Mathematics in Industry"
  http://www.oecd.org/science/sci-tech/41019441.pdf
- Materials Genome Initiative for Global Competitiveness, the White House
  http://www.whitehouse.gov/mgi
- PIRE 2011-Science at the triple point between mathematics, mechanics and materials science
  http://www.math.cmu.edu/PIRE/index.html

References

[AC]   S.M. Allen, J.W. Cahn, Ground state structures in ordered binary alloys with second neighbor interactions. Acta Met. **20**, 423–433 (1972)

[AJ]   A. Avila, S. Jitomirskaya, The ten martini problem. Ann. Math. **170**, 303–342 (2009)

[AZ]   A. Altland, M.R. Zirnbauer, Nonstandard symmetry classes in mesoscopic normal-superconducting hybrid structures. Phys. Rev. B **55**, 1142–1161 (1997)

[Ba]   M. Baake, R.V. Moody (eds.), *Directions in Mathematical Quasicrystals*, CRM Monograph Series 13 (American Mathematical Society, Providence, 2000)

[Bel]   [Bel] B.P. Belousov, A periodic reaction and its mechanism. Sb. Ref. Radiat. Med. (in Russian), (Medzig, Moscow, 1958), pp. 145–147

[Bel94]   J. Bellissard, Lipschitz continuity of gap boundaries for hofstadter-like spectra. Commun. Math. Phys. **160**, 599–613 (1994)

[Bel86]   J. Bellissard, K-Theory of C^*-algebra in solid state physics, in Statistical mechanics and field theory, mathematical aspects, eds by T.C. Dorlas, M.N. Hugenholtzm, M. Winnink, Lecture Notes in Physics. **257** (Springer, New York, 1986), pp. 99–156

[BerHZ]   B. Bernevig, T. Hughes, S.C. Zhang, Quantum spin Hall effect and topological phase transition in HgTe quantum wells. Science **314**, 1757 (2006)

[BES]   J. Bellissard, A. van Elst, H. Schulz-Baldes, The noncommutative geometry of the quantum Hall effect. J. Math. Phys. **35**, 5373–5451 (1994)

[BHZ]   J. Bellissard, D. Herrmann, M. Zarrouati, Hull of aperiodic solids and gap labelling theorems, in Directions in Mathematical Quasicrystals, M.B. Baake, R.V. Moody (eds.), CRM Monograph Series **13**, AMS, Providence (2000), pp. 207–259

[BK]   G. Berkolaiko, P. Kuchment, Introduction to quantum graphs, Mathematical surveys and monographs **186**, American Mathematical Society, Providence (2013)

[BKO]   A. Brataas, A.D. Kent, H. Ohno, Current-induced torques in magnetic materials. Nature Mater. **11**, 372–381 (2012)

[BM86]   J.G. Bednorz, K.A. Müller, Possible high T_c superconductivity in the Ba-La-Cu-O system. Z. fur Phys. B Condens. Matter **64**, 189–193 (1986)

[BM87]   J.G. Bednorz, K.A. Müller, The new approach to high T_c superconductivity. Nobel lecture (1987)

[Bra]   [Bra] K.A. Brakke, *The Motion of a Surface by Its Mean Curvature* (Princeton University Press, New Jersey, 1978)

[CGG]   Y.G. Chen, Y. Giga, S. Goto, Uniqueness and existence of viscosity solutions of generalized mean curvature flow equations. J. Diff. Geom. **33**, 744–786 (1991)

[CH]   J.W. Cahn, J.E. Hilliard, Free energy of a nonuniform system. I. Interfacial free energy. J. Chem. Phys. **28**, 258–267 (1958)

[Con]   A. Connes, *Noncommutative Geometry* (Academic Press Inc, San Diego, 1994)

[Cor63]   A.M. Cormack, Representation of a function by its line integrals, with some radiological applications. J. Appl. Phys. **34**, 2722–2727 (1963)

[Cor64]   A.M. Cormack, Representation of a function by its line integrals, with some radiological applications. II. J. Appl. Phys. **35**, 2908–2913 (1964)

[CR]   R. Caflisch, C. Ratsch, Level set methods for simulation of thin film growth. in Yip, S. (ed.), Handbook of Materials Modeling. vol. I: Methods and Models (Springer, Netherland, 2005), pp. 1–14

[Cra]   J. Crank, *The Mathematics of Diffusion*, 2nd edn. (Oxford University Press, Oxford, 1975)

[Cro]   P.R. Cromwell, *Polyhedra : One of the Most Charming Chapters of Geometry* (Cambridge University Press, Cambridge, 1997)

[CT]   J.W. Cooley, J.W. Tukey, An algorithm for the machine calculation of complex Fourier series. Math. Computation. **19**, 297–301 (1965)

[deB]   N. de Bruijn, Algebraic theory of Penrose's non-periodic tilings of the plane I. Proc. II. Indag. Math. **84**(1), 39–52 (1981)

[DKS]   R.L. Dobrushin, R. Kotecký, S. Shlosman, *Wulff Construction: A Global Shape from Local Interaction*, vol. 104, AMS Translation Series (American Mathematical Society, Providence, 1992)

[EKKST]   P. Exner, J.P. Keating, P. Kuchment, T. Sunada, A. Teplyaev (eds.). Analysis on graphs and its applications. Proc. Symp. Pure Math. **77** (American Mathematical Society, RI, 2008)

[FHKKMS]   M. Furuta, S. Hayashi, M. Kotani, Y. Kubota, S. Matsuo, K. Sato, Bulk-edge correspondence and the Gysin map in K-theory. Preprint

[FK]   L. Fu, C. Kane, Topological insulators in three dimensions. Phys. Rev. B **76**, 045302 (2007)

[Gei]   A. Geim, Random walk to graphene (Nobel lecture). Rev. Mod. Phys. **83**, 851–862 (2011)

[GG]   Y. Giga, S. Goto, Motion of hypersurfaces and geometric equations. J. Math. Soc. Japan **44**, 99–111 (1992)

[GK]   Y. Giga, T. Kobayashi (eds.), What Mathematics Can Do for You: Essays and Tips from Japanese Industry Leaders. (Springer, Japan, 2013)

[Hai]   M. Hairer, Solving the KPZ equation. Ann. Math. **178**(2), 559–664 (2013)

[Hal]   B.I. Halperin, Quantized Hall conductance, current-carrying edge states, and the existence of extended states in a two-dimensional disordered potential. Phys. Rev. B **25**, 2185–2190 (1982)

[Hap]   P.G. Harper, Single band motion of conduction electrons in a uniform magnetic field. Proc. Phys. Soc. Lond. A **68**, 874–892 (1955)

[Hat]   Y. Hatsugai, The Chern number and edge states in the integer quantum hall effect. Phy. Rev. Lett. **71**, 3697–3700 (1993)

[HCO]   S.T. Hyde, L. de Campo, C. Oguey, Tricontinuous mesophases of balanced three-arm, star polyphiles. Soft Matter **5**, 2782–2794 (2009)

[HK]   M.Z. Hasan, C.L. Kane, Colloquium: topological insulators. Rev. Mod. Phys. **82**, 3045–3067 (2010)

[HO]   S.T. Hyde, C. Oguey, Hyperbolic 2D forests and euclidean entangled thickets. Eur. Phys. J. B **16**, 613–630 (2000)

[Hof]   D.R. Hofstadter, Energy levels and wave functions of Bloch electron in rational or irrational magnetic field. Phys. Rev. B **14**, 2239–2249 (1976)

[HOP]   S.T. Hyde, M. O'Keeffe, D.M. Proserpio, A short history of an elusive yet ubiquitous structure in chemistry, materials, and mathematics. Angew. Chem. Int. Ed. **47**, 7996–8000 (2008)

[Hou]   G.N. Hounsfield, Computerized transverse axial scanning (tomography): part 1. Descr. Syst. Br. J. Radiol. **46**, 1016–1022 (1973)

[HR]   S.T. Hyde, S. Ramsden, Polycontinuous morphologies and interwoven helical networks. Europhys. Lett. **50**, 135–141 (2000)

[Hsi]   D. Hsieh et al., A topological Dirac insulator in a quantum spin Hall phase. Nature **452**, 970–974 (2008)

[HYZ]   Y. Han, D. Zhang, L.L. Chng, J. Sun, L. Zhao, X. Zou, J.Y. Ying, A tri-continuous mesoporous material with a silica pore wall following a hexagonal minimal surface. Nat. Chem. **166**, 123–127 (2009)

[Iij]   S. Iijima, Helical microtubules of graphitic carbon. Nature **354**, 56–58 (1991)

[IIK94]   M. Ichikawa, S. Ikeda, Y. Komukai, Measurement of the phase ratio of M_3 to M_1 of alite and its influence on strength development. Cem. Sci. Concr. Technol. (Japan Cement Association) **48**, 76–81 (1994)

[IIK95]   M. Ichikawa, S. Ikeda, Y. Komukai, Estimation of clinker cooling rate by XRD pattern decomposition of ferrite phase and its correlation with strength development. Cem. Sci. Concr. Technol. (Japan Cement Association) **49**, 8–13 (1995)

[IKNSKA]   M. Itoh, M. Kotani, H. Naito, T. Sunada, Y. Kawazoe, T. Adschiri, New Metallic Carbon Crystal. Phys. Rev. Lett. **102**, 055703 (2009)

[INN]   S. Ikeda, T. Nakano, Y. Nakashima, Three-dimensional study on the interconnection and shape of crystals in a graphic granite by X-ray CT and image analysis. Mineral. Mag. **64**, 945–959 (1999)

[IUC]   International Union of Crystallography, Report of the Executive Committee for, Acta Cryst. A **48**(1992), 922–946 (1991)

[JJMW]   J. Joannopoulos, S. Johnsonm, R. Meade, J. Winn, *Photonic Crystals, Modeling the Flow of Light* (Princeton University Press, Princeton, 2008)

[Kaj]   Y. Kajiwara, K. Harii, S. Takahashi, J. Ohe, K. Uchida, M. Mizuguchi, H. Umezawa, H. Kawai, K. Ando, K. Takanashi, S. Maekawa, E. Saitoh, Transmission of electrical signals by spin-wave interconversion in a magnetic insulator. Nature **464**, 262–266 (2010)

[KCND]   H. Kudo, M. Courdurier, F. Noo, M. Defrise, Tiny a priori knowledge solves the interior problem in computed tomography. Phys. Med. Biol. **53**, 2207–2231 (2008)

[KDP]   K. von Klitzing, G. Dorda, M. Pepper, New method for high-accuracy determination of the fine-structure constant based on quantized Hall resistance. Phys. Rev. Lett. **45**, 494–497 (1980)

[Kel]   J. Kellendonk, Noncomutative Geometry of tilings and gap labelling. Rev. Math. Phys. **7**, 1133–1180 (1995)

[KHOCS] H.W. Kroto, J.R. Heath, S.C. Obrien, R.F. Curl, R.E. Smalley, C60, Buckminsterfullerene. Nature **318**, 162–163 (1985)

[Kita]   A. Kitaev, Periodic table for topological insulators and superconductors. arXiv:0901.2686

[Kitt]   C. Kittel, *Introduction to Solid State Physics*, 8th edn. (Wiley, New York, 2004)

[KL]   C. Kipnis, C. Landim, *Scaling Limits of Particle Systems* (Springer, New York, 1999)

[KM]   C.L. Kane, E.J. Mele, Quantum spin Hall effect in graphene. Phys. Rev. Lett. **95**, 226801 (2005)

[Kob93] R. Kobayashi, Modeling and numerical simulations of dendtitic crystal growth. Phys. D **63**, 410–423 (1993)

[Kob94] R. Kobayashi, A numerical approach to three-dimensional dendritic solidification. Exp. Math. **3**, 59–81 (1994)

[Kön]   M. König, S. Wiedmann, C. Brüne, A. Roth, H. Buhmann, L.W. Molenkamp, X.-L. Qi, S.-C. Zhang, Quantum spin Hall insulator state in HgTe quantum wells. Science **318**, 766–770 (2007)

[Kot02] M. Kotani, A central limit theorem for magnetic transition operators on a crystal lattice. J. Lond. Math. Soc. **65**, 464–482 (2002)

[Kot03] M. Kotani, Lipschitz continuity of the spectra of the magnetic transition operators on a crystal lattice. J. Geom. Phys. **46**, 323–342 (2003)

[KP]   J. Kellendonk and I.F. Putnam, Tilings, C^*-algebras, and K-theory, in Directions in Mathematical Quasicrystals. CRM Monograph Series13, M. Baake and R.V. Moody (eds.) (2000)

[KRS]   J. Kellendonk, T. Richter, H. Schulz-Baldes, Edge current channels and Chern numbers in the integer quantum Hall effect. Rev. Math. Phys. **14**, 87–119 (2002)

[KS00]   M. Kotani, T. Sunda, Albanese maps and off diagonal long time asymptotics for the heat kernel. Comm. Math. Phys. **209**, 633–670 (2000)

[KS01]   M. Kotani, T. Sunada, Standard realizations of crystal lattices via harmonic maps. Trans. Amer. Math. Soc. **353**, 1–20 (2001)

[KS02]   M. Kotani, T. Sunada, Spectral geometry of crystal lattices. Heat kernels and analysis on manifolds, graphs, and metric spaces (Paris, 2002), Contemp. Math. vol. 338, American Mathematical Society Providence, 2003) pp. 271–305

[KSR]   H. Kudo, T. Suzuki, E.A. Rashed, Image reconstruction for sparse-view CT and interior CT -introduction to compressed sensing and differentiated backprojection. Quant. Imaging. Med. Surg. **3**, 147–161 (2013)

[KSV]   M. Kotani, H. Schulz-Baldes, C. Villegas-Blas, Quantization of interface currents. J. Math. Phys. **55**, 121901 (2014)

[Kub]   Y. Kubota, Bulk-edge correspondence via coarse geometry. Preprint

[Kuc]   P. Kuchment, The mathemtics of photonic crystals, in Mathematical Modeling in
           Optical Science, G. Bao, L. Cowsar, W. Master (eds.), Front. Appl. Math. **22**, SIAM
           (2001), pp. 207–272

[KWHH]   Y. Kamihara, T. Watanabe, M. Hirano, H. Hosono, Iron-based layered superconductor
           La[$O_{1-x}F_x$]FeAs (x = 0.05−0.12) with T_c = 26 K. J. Am. Chem. Soc. **130**, 3296–3297
           (2008)

[KZ]   M. Kardar, Y.-C. Zhang, Scaling of directed polymers in random media. Phys. Rev.
           Lett. **58**, 2087–2090 (1987)

[LER]   T. Lery, M. Primicerio, M.J. Esteban, M. Fontes, Y. Maday, V. Mehrmann, G. Quadros,
           W. Schilders, A. Schuppert, H. Tewkesbury, *European Success Stories in Industrial
           Mathematics* (Springer, Berlin, 2012)

[Lu]   H. Lüth, *Surfaces and Interfaces of Solids* (Springer, New York, 1993)

[Mac62]   A.L. Mackay, A dense non-crystallographic packing of equal spheres. Acta Cryst. **15**,
           916 (1962)

[Mac81]   A.L. Mackay, De Nive Quinquangula. Krystallografiya **26**, 910–918 (1981)

[Mac82]   A.L. Mackay, Crystallography and the Penrose pattern. Phys. A **114**, 609–613 (1982)

[Mar]   D.W. Marquardt, An algorithm for least-squares sstimation of nonlinear parameters.
           J. Soc. Ind. Appl. Math. **11**, 431–441 (1963)

[MB]   J. Moore, L. Balents, Topological invariants of time-reversal-invariant band structures.
           Phys. Rev. B **75**, 121306 (2007)

[MKWL]   Y.W. Mo, J. Kleiner, M.B. Webb, M.G. Lagally, Activation energy for surface diffusion
           of Si on Si(001): a scanning-tunneling-microscopy study. Phys. Rev. Lett. **66**, 1998–
           2001 (1991)

[MNZ04]   S. Murakami, N. Nagaosa, S.-C. Zhang, Phys. Rev. Lett. **93**, 156804 (2004)

[Mor]   C.B. Morrey, On the derivation of the equatiions of hydrodymanics from statistical
           mechanics. Comm. Pure Appl. Math. **8**, 279–326 (1955)

[MRPC]   K. Morgenstern, G. Rosenfeld, B. Poelsema, G. Comsa, Brownian motion of vacancy
           islands on Ag(111). Phys. Rev. Lett. **74**, 2058–2061 (1995)

[MSSL]   R. McGrath, H.R. Sharma, J.A. Smerdon, J. Ledieu, The memory of surfaces: epitaxial
           growth on quasi-crystals. Phil. Trans. Roy. Soc. A **370**, 2930–2948 (2012)

[Mul]   W.W. Mullins, *The motion of a surface by its mean curvature* (Princeton University
           Press, New Jersey, 1978)

[N]   Y. Nishiura, Far-from-equilibrium dynamics, translations of mathematical mono-
           graphs. Amer. Math. Soc. **209**, 311 (2002)

[Nak]   S. Nakashima, Diffusivity of ions in pore water as a quantitative basis for rock defor-
           mation rate estimates. Tectonophysics **245**, 185–203 (1995)

[Nat]   Nature Editorial Office, Nature Milestones in Crystallography. Nature 11th supple-
           ment (2014)

[Nis]   T. Nishinaga (ed.), Handbook of Crystal Growth, 2nd (edn.) Fundamentals: thermo-
           dynamics and kinetics, vol. IA. Elsevier (2014)

[NP]   G. Nicolis, I. Prigogine, *elf-Organization in Nonequilibrium Systems—From Dissipa-
           tive Structures to Order through Fluctuations* (Wiley, New York, 1977)

[Nov]   K. Novoselov, Materials in the flatland (Nobel lecture). Rev. Mod. Phys. **83**, 837–849
           (2011)

[NSO]   K. Nagata, S. Sugita, M. Okada, Bayesian spectral deconvolution with the exchange
           Monte Carlo method. Neural Netw. **28**, 82–89 (2012)

[OJK]   T. Ohta, D. Jasnow, K. Kawasaki, Universal scaling in the motion of random interfaces.
           Phys. Rev. Lett. **49**, 1223–1226 (1982)

[OS]   S. Osher, A. Sethian, Fronts propergating with curvature dependent speed, algorithms
           based on Hamiltonian-Jacobi formulations. J. Comp. Phys. **79**, 12–49 (1988)

[Osa]   E. Osawa, Superaromaticity. Kagaku **25**, 854–863 (1970)

[PBNZ]   S. Pratontepa, M. Brinkmanna, F. Nüeschb, L. Zuppirolib, Nucleation and growth of
           ultrathin pentacene films on silicon dioxide: effect of deposition rate and substrate
           temperature. Synth. Met. **146**, 387–391 (2004)

[Pen]   R. Penrose, The rôle of aesthetics in pure and applied mathematical research. Bulletin of the Institute of Mathematics and Its Applications **10**, 266–271 (1974)

[SBGC]   D. Shechtman, I. Blech, D. Gratias, J. Cahn, Metallic phase with long-range orientational order and no translational symmetry. Phys. Rev. Lett. **53**, 1951–1953 (1984)

[Sci]   Science Crystallography 100. **343**, 1091–1116 (2014)

[Sen]   M. Senechal, *Quasicrystals and Geometry* (Cambridge University Press, Cambridge, 1995)

[SS]   J.T. Sadowski, G. Sazaki, S. Nishikata, A. Al-Mahboob, Y. Fujikawa, K. Nakajima, R.M. Tromp, T. Sakurai, Single-nucleus polycrystallization in thin film epitaxial growth. Phys. Rev. Lett. **98**, 046104 (2007)

[Su94]   T. Sunada, A discrete analogue of periodic magnetic Schrödinger operators. Contemporary Math. **173**, 283–299 (1994)

[Su08]   T. Sunada, Crystals that nature might miss creating. Not. Am. Math. Soc. **55**, 208–215 (2008)

[Su13]   T. Sunada, Topological Crystallography, Survey and tutorials in the Applied Mathematical Sciences, vol. 6 (Springer, Japan, 2013)

[Sun57]   I. Sunagawa, Variations in crystal habit of pyrite. Rep., Geol. Surv. Jpn **175**, 1–42 (1957)

[Sun05]   I. Sunagawa, *Crystals: Growth, Morphology, and Perfection* (Cambridge University Press, Cambridge, 2005)

[RT]   S. Ryu, T. Takayanagi, Topological insulators and superconductors from D-brane. Phys. Lett. B **693**, 175–179 (2010)

[Tay]   J.E. Taylor, The structure of singularities in soap-bubble-like and soap-film-like minimal surfaces. Ann. Math., Second Ser. **103**, 89–539 (1976)

[TIM]   A.-P. Tsai, A. Inoue, T. Masumoto, A stable quasicrystal in Al-Cu-Fe system. Jpn. J. Appl. Phys. **26**, L1505–L1507 (1987)

[TKNN]   D.J. Thouless, M. Kohmoto, P. Nightingale, M. den Nijs, Quantized Hall conductance in a two-dimensional periodic potential. Phys. Rev. Lett. **49**, 405–408 (1982)

[TLNKK]   M. Tagami, Y. Liang, H. Naito, Y. Kawazoe, M. Kotani, Negatively curved cubic carbon crystals with octahedral symmetry. Carbon **76**, 266–274 (2014)

[Tor]   H. Toraya, Array-type universal profile function for powder pattern fitting. J. Appl. Cryst. **23**, 485–491 (1990)

[Tur]   A.M. Turing, The chemical basis of morphogenesis. Phil. Trans. Roy. Soc. Lond. B **237**, 37–72 (1952)

[WCBET]   J.-M. Wen, S.-L. Chang, J.W. Burnett, J.W. Evans, P.A. Thiel, Diffusion of large two-dimensional Ag clusters on Ag(100). Phys. Rev. Lett. **73**, 2591–2594 (1994)

[YTNK]   H. Yanagisawa, T. Tamaki, M. Nakamura, K. Kudo, Structural and electrical characterization of pentacene films on SiO_2 grown by molecular beam deposition. Thin Solid Films **464–465**, 398–402 (2004)

[Zha]   A.M. Zhabotinsky, Periodic processes of the oxidation of malonic acid in solution. Bipfizika **9**, 306–311 (1964)

Chapter 3
Some Specific Examples
of Mathematics–Materials Science
Collaboration at AIMR

Abstract In today's modern research environment, the emphasis is placed on the direct interaction of mathematics with other scientific fields and industry groups to solve complex problems jointly faced by various disciplines. The demand for the participation of mathematicians in materials science is growing. One reason for this growth is to provide new mathematical tools to understand the relationship between microscopic geometric structures and macroscopic properties and functions. Although much technology for observations at the atom/molecule level has largely been developed, we need mathematical tools to analyze and interpret the experimental data in a deeper and more sophisticated way and to create a conceptual understanding of the essence of microscopic structures. Another reason is to establish a basis for the prediction and smart design of novel functional materials by using accumulated data and high performance computatsion. Of course, these two factors are related. At AIMR, we set up a place where world-leading materials scientists (both experimentalists and theorists) and mathematicians meet daily and discuss materials science problems to establish new science. Here we present some of the results that have emerged.

Keywords Computational homology · Stochastic model · Carbon nanotube · Mackay crystal · Graph theory

In this chapter, we propose new directions for collaborations between mathematics and materials science, and present some of the results emerging along these lines. We believe the following three themes are promising because they are important issues in materials research requiring new mathematical frameworks to their resolution.

Target 1: Non-equilibrium Materials Based on Mathematical Dynamical Systems

One of the major challenges in materials science is to synthesize materials, in which multi-functionality emerges based on non-equilibrium states, and hybrid structures consisting of different types of materials or non-homogeneous systems. From the perspective of mathematical dynamical systems, we shall focus on understanding the mechanisms associated with the formulation of dynamical structures in

© The Author(s) 2015
S. Ikeda and M. Kotani, *A New Direction in Mathematics
for Materials Science*, SpringerBriefs in the Mathematics of Materials,
DOI 10.1007/978-4-431-55864-4_3

non-equilibrium systems. This will enable us to accurately control non-equilibrium and inhomogeneous materials and to achieve prescribed multi-functionalities under a given environment. This project targets, for example, metallic glasses, polymer glasses, block copolymers, bio-inspired materials, and super-hybrid multifunctional devices for a green society.

Target 2: Topological Functional Materials

Topology is a mathematical concept for describing a shape robust against continuous deformation. It supplies tools to abstract essential properties from a complex shape and deform it into a simpler shape. One challenge in materials science using topology is to synthesize functional materials that are robust under environmental change but achieve highly sensitive responses to inputs at the same time. This project targets, for example, spintronics materials, superconductors, and MEMS devices for energy-saving, along with nanoporous metal catalysts and new materials for photo-voltaic solar energy conversion and thermoelectric conversion for energy-harvesting.

Target 3: Multi-Scale Hierarchical Materials Based on Discrete Geometric Analysis

Innovative functional materials can be created only by recognizing the complex multi-scale hierarchical structure in materials systems from the atomic/molecular scale to the macroscopic scale of materials and devices. Therefore, understanding and exploiting multi-scale hierarchies form a part of the fundamental research performed at AIMR. Precise structure analysis and control at each level of hierarchy from bottom up will be carried out using state-of-the-art equipment and latest technologies.

In addition to experimental technologies, AIMR will apply mathematical methods such as discrete geometric analysis to this hierarchical problem. This analysis bridges the various scales taking in account detailed geometric data. By employing these advanced tools, we are attempting to produce functional multi-scale hierarchical materials. This project targets the identification of mid-range and long-range order in the atomic arrangement of bulk metallic glasses and interfacial processes from the atomic/molecular level to macroscopic properties. Grain boundaries are one example where improvements to electrical conduction in devices can be made. Another example is solid–liquid interfaces where improved control can eliminate friction, thereby saving energy.

The following presents concrete results obtained over the last 3 years from research within the target projects described above for the mathematics–materials science collaboration.

3.1 Elucidation of Metallic Glass Structure by Computational Homology

A. Hirata et al., Science **341** *(2013) 376–379.* [Hir]

The term *metallic glass* is often confused with the term *amorphous metal*. Amorphous metals are simply metallic materials that have random atomic arrangement. The term does not involve any other attribute such as a process of formation. Metallic glass is an amorphous metal that possesses a **glass transition**, similar to silica-based glasses (window glasses); that is, there is a supercooled condition between the melting point and glass transition temperature, in which metallic glass exhibits a viscous or molten state between solid and liquid, which enables it to be molded into any shape to make glasswork. Such glasses have two benefits, amorphicity and malleability.

Metallic materials have the tendency to form crystalline structures; the atoms in the molten metal arrange themselves into a regular structure very quickly when they are cooled below the liquidus temperature. In some metallic systems, composed of specifically three or four kinds of elements, this quick rearrangement of atoms is suppressed so that bulk metallic glasses are formed. Metallic glasses display extremely good physical, chemical, and mechanical properties. For example, some types of metallic glasses are several times stronger than ordinary steel and this is the reason why metallic glass attracts much attention. Metallic glass does not have any grain boundaries, the sources of fractures in other metallic materials, and this is the main reason for its excellent properties.

As defined above, the structure of metallic glass is assumed to be almost random and does not have any short-, mid-, or long-range order. Nevertheless, to explain why atoms are prevented from forming crystalline structures and instead form metallic glass, one hypothesizes that atomic clusters form icosahedra and this prevents crystalline structures to form because icosahedral packing of space is prohibited. This implies that there is some order in the disordered (random) structure that resolves the longtime mystery. The description of the global structure is a big challenge in both materials science and mathematics.

The research team of mathematicians and experimentalists at AIMR has successfully characterized the atomic structure in metallic glass and explained the relationship between atomic-scale ordering (in random structure) and its ability to form glass.

First, the experimentalists developed an angstrom beam electron diffraction method, analyzing the local atomic structure (Fig. 3.1). Although, with a thick electron beam, the diffraction pattern gave a Debye–Scherrer ring indicating random structure, some characteristic structures appeared when the diameter of the beam decreased to several Angstroms.

Mathematicians applied "computational homology" to analyze the observed structure (Fig. 3.1). Computational homology gives Betti numbers $b_0, b_1, b_2, \ldots, b_n$, which can be used to distinguish topological spaces based on the connectivity of the n-dimensional simplicial complexes. The collaboration unveiled a long-standing mystery related to the atomic configuration: geometric distortions of icosahedral

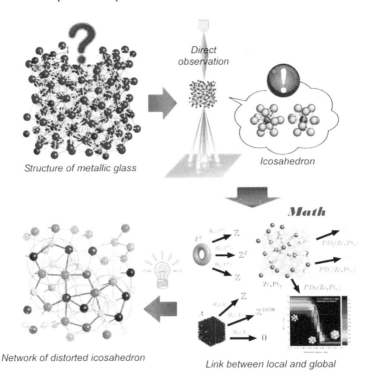

Fig. 3.1 Structure of metallic glass from studies using electron beam diffraction and computational homology. Courtesy of Dr. Akihiko Hirata, Tohoku University

clusters in metallic glass can be scaled up to produce long-range disorder with topological connectivity. The co-existence of icosahedral and face-centered cubic (fcc)-like symmetries in the distorted icosahedral clusters leads to the perfect distortion for making a disordered, densely-packed structure. The underlying discrepancy that remained unknown for half a century has been resolved using *persistent homology*. The details will be discussed in the second volume of this series SBMM.

A mathematical study of metallic glasses has just started. The description of mid-range and long-range order (some results appear in [NHHEMN] using computational homology and persistent diagrams) will be followed by the elucidation of material properties, mechanisms in the formation of glass, and the glass transition phenomena.

Mathematical Methods and Tools Used in this Study:

Computational persistent homology
Software CHomP has become standard in computational homology. URL: http://chomp.rutgers.edu/

3.2 Application of a Stochastic Model

Stochastic modeling is useful in materials research because physical and chemical phenomena are often controlled by probabilistic processes. In this section, we present two examples for which stochastic modeling was applied to actual materials, one being the collision of atoms during molecular beam epitaxy and the other being the release of stress due to the formation of share bands during deformation.

3.2.1 Stoichiometry Control Based on a Mathematical Model

D.M. Packwood et al., Phys. Rev. Lett. **111** *(2013) 036101.* [PSH]

Thin films are among those indispensable materials used for quantum electronic devices, such as transistors, light emitting diodes (LEDs), and lasers. The quality of thin films, that is, structure (e.g., thickness, flatness, density of defects) and chemical composition (e.g., stoichiometry, concentration of impurities) are important in obtaining the appropriate quantum properties of devices. The usual technique to make high-quality thin films is either molecular beam epitaxy (MBE) or molecular beam deposition (MBD). Using an ultra-high vacuum chamber, MBE produces epitaxial films grown on single-crystalline substrates (with a specific orientation between substrate and thin film), whereas, using the same equipment, MBD deposits thin films on substrates that are not single crystals. Although resistance heating of effusion cells is usually used in the evaporation of source materials, oxide materials are not easy to evaporate because of their high melting temperature. Pulsed laser deposition (PLD) is a promising variant of MBE that is used to make oxide thin films. Laser irradiation easily evaporates high-melting-temperature materials and hence enables oxide thin films to be produced. One of the problems of PLD is that the stoichiometry of the thin film often differs from the target surface because of differences in behavior between elements (e.g., Li and Mn in this study) during their evaporation from source materials, the nature of the plume dynamics, and the manner in which the Li and Mn ions deposit onto the substrate surface. To prevent changes in stoichiometry, for example, the oxygen partial pressure is controlled during deposition. However, the detailed microscopic mechanism leading to nonstoichiometry remains unknown and a way to improve the situation is unclear.

In this study, a team of a theoretical chemist and experimental materials scientists at AIMR succeeded in developing a promising analytic model to quantitatively describe the cation nonstoichiometry in PLD of oxide films. The main target material in this study was lithium manganate, which is used to develop high-capacity lithium-ion batteries. Their model is the first analytic model of collision-induced plume expansion that can predict the partial oxygen pressure dependence of the Li content of a thin film. The model assumes that, after being ejected from the target surface under laser ablation, the lithium (Li) or manganese (Mn) atoms undergo a sequence of elastic collisions with the surrounding O_2 molecules. 'Elastic' means that during a collision, the total kinetic energy of the atom (Li or Mn) and the O_2 molecule is conserved, i.e., that no energy is lost to internal modes of the O_2 molecule. Hence, the kinetic energy of the atom can change as a result of collision with an O_2 molecule, providing that this change is properly compensated by a change in kinetic energy of the O_2 molecule. Each collision is assumed to be a random event independent of any other; moreover, collisions with either Li or Mn atoms in the plume are neglected. Based on these assumptions, a stochastic differential equation describing the trajectory of the atom (Li or Mn) was derived, specifically

$$\frac{d\mathbf{R}_t}{dt} = c^{N_t}\mathbf{V}_0 + b\sum_{k=1}^{N_t} c^{N_t-k}\mathbf{W}_k, \tag{3.1}$$

where $\mathbf{R}_t$ is the position of one Li or Mn in the plume at time t, $\mathbf{V}_0$ the ejection velocity of the atom from the surface at time 0, and b and c are constants that depend upon the mass ratio of the Li or Mn atoms and the mass of the O_2 molecule, $\mathbf{W}_k$ is the velocity of the kth oxygen molecule that collides with the Li atom and N_t is the number of collisions that have occurred by time t. Moreover, N_t parameterizes the Poisson process, and $\mathbf{W}_i$, $(i = 1, 2, \ldots)$ are independent and identically distributed random variables for the Maxwell distribution. Equation (3.1) cannot be solved analytically, however it is fairly easy to integrate numerically; representative trajectories of the Li and Mn atoms in a chamber at 298 K containing 10^{-6} torr of O_2 are shown in Fig. 3.2a. Whereas Eq. (3.1) cannot be solved analytically for $\mathbf{R}_t$, the two-dimensional probability density function $f_a(r_x, r_y, t)$ can be computed analytically, although the resulting formula is too long to present here. The quantity $f_a(r_x, r_y, t)dr_x dr_y$ is understood as the probability of finding an atom of type a (Li or Mn) at point (r_x, r_y) at time t. Assuming a substrate of width $2L_s$ perpendicular to the target surface at height H above it, the ratio of Li to Mn landing on the substrate surface can be computed as

$$\rho = \frac{\int_0^\infty dt \int_{-L_s}^{L_s} dr_x f_{\text{Li}}(r_x, H; t)}{\int_0^\infty dt \int_{-L_s}^{L_s} dr_x f_{\text{Mn}}(r_x, H; t)}. \tag{3.2}$$

We compared the predictions of Eq. (3.2) with experimental data obtained using a $Li_{1.3}Mn_2O_2$ target surface, and obtained good agreement between the two (Fig. 3.2b). The origin of this difference can be traced to the trajectory simulations in Fig. 3.2, in which the light Li atom is scattered much more violently by the O_2 molecules than the relatively heavy Mn atom. This result demonstrates that in PLD of targets containing light and heavy elements, there inevitably is a stoichiometric reduction in the amount of light element because of differences in scattering of the two elements. The best route to control the stoichiometric ratio of elements in a thin film is therefore expected to be through careful control of the O_2 pressure. This point was not so obvious before this study was performed.

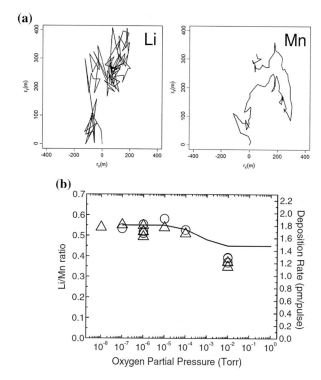

Fig. 3.2 **a** Simulated trajectories of a Li atom and a Mn atom. **b** Oxygen partial pressure dependence of Li/Mn ratio of the thin film. Line is the prediction from Eq. (3.2). Reprinted figures with permission from [PSH] (http://dx.doi.org/10.1103/PhysRevLett.111.036101). Copyright (2013) by the American Physical Society

Mathematical Methods and Tools Used in this Study:

Stochastic processes and probability theory.

3.2.2   Deformation Analysis of Bulk Metallic Glass Using a Stochastic Model

D.V. Louzguine et al., Journal of alloys and compounds **561** *(2013) 241-246.* [LPXY]
    Metals are typical materials that show shear bands when they are deformed by stress. The application of continuous stress to a solid causes an increase in strain. However, when shear bands appear, the applied stress temporarily decreases because of the temporal release from the stressed condition. Although this decrease occurs periodically, the periodicity cannot be predicted deterministically and appears to be stochastic in nature. If we can establish a complete theory for this stochastic

process, then the deformation behavior of metallic materials including bulk metallic glass (BMG) will have greater predictive power, as well as design capability of new materials.

In this study, a team of experimental materials scientists and a theoretical chemist analyzed the deformation of BMG. They produced a BMG of $Ni_{50}Pd_{30}P_{20}$ using a flux treatment and casing, and confirmed the formation of a glassy structure by X-ray diffraction and transmission electron microscopy. Its deformation behavior under uniaxial compression at room temperature was studied at the strain rate of $5 \times 10^{-4}s^{-1}$, as well as at three strain rates at quasistatic loading conditions. A chaotic jagged curve is observed in the stress-strain diagram shown in Fig. 3.3a. In compression experiments, the share bands can clearly be seen by microscopic observation. The distribution of maximum stress before each stress drop (Fig. 3.3b) is not a simple Gaussian distribution, indicating the complexity of this phenomenon.

The team introduced a stochastic model to deal with the chaotic behavior. The model studies the probability distribution of the intervals $U_2 - U_1$, $U_3 - U_2$,... between successive stress drops. It assumes that, if a stress drop occurs at time t, then the average length of time until a new shear band is a constant $1/\lambda_1$, and the average length of time until an existing shear band propagates further is N_t/λ_0,

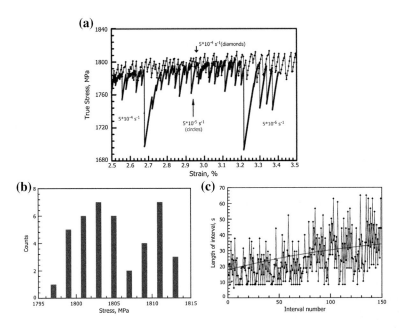

Fig. 3.3 a Calculated (*upper*) and experimentally observed (*lower*) stress-strain curves of the $Ni_{50}Pd_{30}P_{20}$ rod tested at $5 \times 10^{-4}s^{-1}$. **b** The distribution of maximum stress before each stress drop. **c** Experimentally observed length of interval. The solid line is the average interval length calculated using Eq. (3.3) with $\lambda_1 = 0.056s^{-1}$, $\lambda_0 = 0.023s^{-1}$. Reprinted from [LPXY] (http://dx.doi.org/10.1016/j.jallcom.2013.01.193), Copyright (2013), with permission from Elsevier

where N_t is the number of shear bands present on the rod at time t. Based on these assumptions, it was found that the average of the intervals $U_2 - U_1, U_3 - U_2, \ldots$ is

$$E(U_k - U_{k-1}) = \frac{E(N_{k-1})}{M}(\frac{1}{\lambda_0} - \frac{1}{\lambda_1}) + \frac{1}{\lambda_1}, \qquad (3.3)$$

where M is the number of places on the rod where a shear band may appear. By fitting the average of the data (Fig. 3.3c) using Eq. (3.3), we can estimate the values of λ_1 and λ_0. This analysis suggests that the jagged flow initially results from the appearance of new shear bands in the material. However, the more stress is applied, the nature of these dynamics changes because of the propagation of shear bands that appear in the earlier stage. This study clearly shows the importance of stochastic modeling in characterizing deformations of materials.

Mathematical Methods and Tools Used in this Study:

Stochastic modeling

3.3 New Geometric Measures for Finite Carbon Nanotubes

T. Matsuno et al., Pure and Applied Chemistry **86** *(2014) 489–495.* [MNHSKI].
M. Tagami et al., Carbon **76** *(2014), 266-274* [TLNKK].

Many a proposal and attempts to synthesize carbon materials are made these days from both mathematicians and organic chemists. Unfortunately, there has been little direct collaboration between them, and in most cases chemists look for mathematical explanations only after their success of synthesis, whereas mathematicians propose and classify structures without knowing much of the technology related to synthesis. New relations, however, start in many places and opportunities of encounters increase. Research results presented here strongly suggest that collaboration between mathematicians and synthetic chemists is fruitful in that mathematics provides a vocabulary to describe molecule structures and guiding principles to synthesize stable materials.

Over the past three decades, there have been fascinating discoveries in carbon materials. Fullerene C_{60}, also called Buckminsterfullerene or buckyball, was first discovered in 1985 by Richard Smalley, Robert Curl, and Harold Kroto [KHOCS]. They were awarded the Nobel Prize in Chemistry in 1996. In 1991, Sumio Iijima published his paper in *Nature* [Iij] and reported the discovery of carbon nanotubes (CNTs). After the discoveries of these materials, an amazing number of studies were conducted. Furthermore, still fresh in memories is the discovery by Andre Geim and Konstantin Novoselov of a simple way to prepare a two-dimensional carbon sheet, now called graphene, which won the pair the 2010 Nobel Prize in Physics for their groundbreaking experiments. The characteristic feature of these carbon allotropes is that they are fundamentally made up of six-membered rings of carbon atoms.

Fullerene C_{60} includes also five-membered rings. Carbon materials may also have a small number of seven-membered rings as "defects" that influence their properties.

Based on these geometric characteristics, the structure of carbon materials can be analyzed mathematically. Many attempts at classifying these carbon networks under certain conditions have been described. M. Deza's book [DS], for example, describes the graph-theoretic and geometric properties of fullerenes and their generalizations, specifically planar graphs in which all faces are polygonal with only two possible lengths. Some of the structures were synthesized by chemists without knowledge of this classification.

We relate an example of the collaborative work between organic chemists and mathematicians to find an indexing scheme and guiding principles in designing carbon structures. In their study, synthetic chemists with the help of a mathematician (geometer) investigated new geometric measures specifically for finite CNT molecules. In 1992, a chiral indexing using coordinates (n,m) were proposed for these geometric measures [SFDD] and widely accepted. However, there has been no measure for finite CNT molecules because such molecules did not exist at that stage. In recent years, the synthesis of finite CNTs with discrete sizes has been successful [HNYI], and the need arises for an index to measure length, bond-filling, and atom-filling rates. From the study, the researchers succeeded in obtaining a new geometric index for this purpose (Fig. 3.4). This index hopefully helps in advancing the science and technology of finite CNT molecules [MNHSKI].

In 1991, Mackay and Terrons [MT] proposed a carbon crystal structure, now called the *Mackay-Terrones Schwarzite crystal*, or simply the *Mackay crystal* (Fig. 3.5). They adopted *Gauss curvature* to account for the positively curved surface (i.e., a sphere) of fullerene, the surface of zero curvature (i.e., a plane) of graphene, and pro-

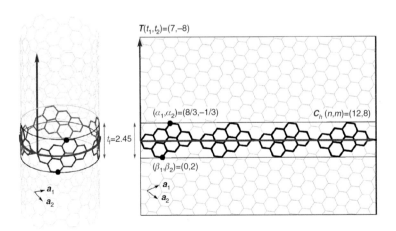

Fig. 3.4 Relationship between the previous descriptor for CNTs proposed by Saito et al. [SFDD] and the new descriptor for finite CNT molecules developed in this study. Reproduced from De Gruyter [MNHSKI], Walter De Gruyter GmbH Berlin Boston, [2014].

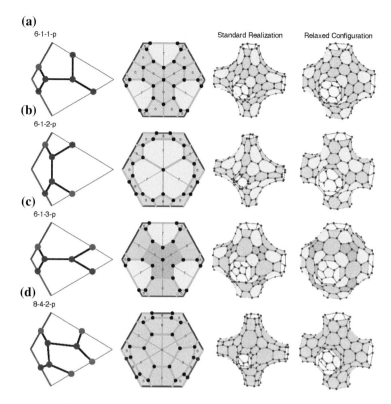

Fig. 3.5 Mackay-like crystals. Reprinted from [TLNKK] (http://dx.doi.org/10.1016/j.carbon. 2014.04.077), Copyright (2014), with permission from Elsevier

posed the Schwarz minimal P-surface, which has a negative Gauss curvature, for a new carbon structure. Because surfaces with negative curvatures are mathematically more stable under local deformation, a search for other negatively curved networks might turn up new structures. One such trail is found in [TLNKK], where a systematic search was performed of triply periodic negatively curved carbon networks of sp^2-bonding and study of their electrical classification (metal, semiconductor or insulator).

From a mathematical point of view, Mackay's idea is attractive but not satisfactory because for a given carbon network, finding a surface on which the network lies is nontrivial. A definition of *curvatures* of a *discrete surface* (carbon networks) is needed. There is a well-known notion of curvature for a polyhedron defined by the angle defect at a vertex, specifically, the difference of the sum of all angles of adjacent edges meeting at a vertex from 2π.

In many cases, including the Mackay crystals, this notion does not work for two reasons. First, the various polygons (hexagons, pentagons, heptagons) do not lie on a flat plane, and second, there is no angle defect which satisfies the conditions for a sp^2 bonding (three coplanar edges meet at a vertex with $120°$ angles between adjacency edges) although the network seems to be a negatively curved discrete surface. A definition of Gauss curvature and mean curvature of a discrete surface will be discussed elsewhere by one of us (MK). These two notions provide useful guidance because the Gauss curvature indicates inner frustrations and the mean curvature external stability.

The Mathematical Methods and Tools Used in this Study:

Graph theory
Combinatrics
Topology
Discrete surface theory

3.4 Materials Having Network Structures

A variety of network structures can be seen in materials, from the atomic/molecular level to the mesoscopic/macroscopic level. Therefore, organizing materials from the viewpoint of network structures is possible. Two network structures seen for a molecular structure of a single molecule and a connected pore structure of nanoporous gold were studied by AIMR researchers using mathematical theories such as graph theory and probability theory. The results are briefly given below.

3.4.1 Mathematical Technique Predicts Molecular Magnet

*D.M. Packwood et al., Proceedings of the Royal Society A**469** (2013) 20130373.* [PRFKT]
    As well as the result of the last Sect. 3.3, the result presented in this section suggests how applied mathematics can contribute to the synthesis of new molecules having novel properties and functions.
    Molecular magnets are cluster molecules containing 10–100 transition-metal ions. The unpaired electron spins on these ions couple through superexchange interactions, giving the molecule a large total spin and other interesting magnetic properties. This unique behavior is attracting much attention for high-density information storage and spintronics-based computing. A prototypical example of a molecular magnet with total spin $S = 10$, $Mn_{12}O_{12}(O_2CCH_3)_{16}$, is shown in Fig. 3.6a. A major problem in this field is that the magnetic properties of molecular magnets tend to be greatly suppressed when these molecules are adsorbed on a surface. The suppression is obtained from redox chemistry between the surface and transition metal ions in the molecule, and also from structural deformations of the molecule upon adsorption. It is extremely important from a technological point-of-view to achieve the adsorption

(a)   **(b)**

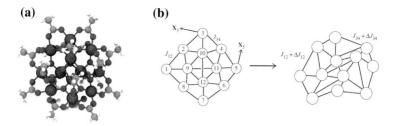

Fig. 3.6 **a** Structure of the Mn$_{12}$ (acetate) molecule. **b** A model of random shape deformation of Mn$_{12}$-like molecular magnet. $\mathbf{X}_k$ is a random Gaussian vector added to point k. Reproduced from [PRFKT] with permission of the Royal Society

of the molecular magnets onto the surfaces with their magnetic properties intact. For this reason, experimental materials scientists who are collaborating with a theoretical chemist are developing innovative "mathematical chemistry" techniques that can predict the structure of novel molecular magnets with magnetic properties that are highly stable under structural deformations.

Regarding the basic model used in this study (Fig. 3.6b, left-hand figure), the transition metal ions are represented as points in a three-dimensional (3D) space, and edges between points are drawn to represent superexchange interactions between the ions. This diagram is referred to as an interaction graph. Physically, the interaction graph corresponds to a spin Hamiltonian,

$$\hat{H} = -\sum_{i,j} J_{ij}\hat{\mathbf{s}}_i \cdot \hat{\mathbf{s}}_j, \qquad (3.4)$$

where the sum runs over all pairs of interacting transition metal ions, $\hat{\mathbf{s}}_i$ is the spin operator for the electron on transition metal i, and J_{ij} the exchange constant between ions i and j. Starting from this 'unperturbed' interaction diagram, we add random Gaussian vectors to the positions of each transition metal ion to model the effect of a structural deformation (Fig. 3.6b, right-hand side). This deformation changes the exchange constant J_{ij} by an amount $c\Delta r_{ij}$, denoting the change in distance between ions i and j under the deformation, and c is a constant that measures the sensitivity of the exchange constants to the deformation.

How can we use this model to predict the properties of molecular magnets that are robust under structural deformation? We let P be a probability measure and ε a small positive number; then we proposed the following criterion:

$$P(R < \varepsilon) > 1 - \varepsilon, \qquad (3.5)$$

where R is the fractional change in the energy of the ground state of the spin, specifically

$$R = \left| \sum_{ij} \Delta J_{ij} / \sum_{ij} J_{ij} \right|. \tag{3.6}$$

If the above criterion is satisfied, then there is a very small probability that the energy of the ground state of the total spin will change under a deformation, and we would not expect transitions to other spin states to occur under these conditions. A molecule that satisfies (3.5) for a small ε is said to have a *weak topological invariant magnetic moment*, for the following reason. R depends upon the connectivity of points in the interaction graph, and according to (3.6), there is a very small probability that R will change under deformation, providing that this deformation does not make or break connections in the interaction graph. This behavior qualifies as a topological invariant. In contrast, some deformations can cause large changes in R, but these have a very small probability (less than ε) of occurring. Under this circumstance, we say the topological invariant is 'weak'. The definition of a weak topological invariant in equation (3.5) is closely related to the Ky–Fan metric for convergence of the probability as used in probability theory.

Through a series of calculations, it was found that molecules in which the transition metal ions lie in a 2D plane and have ferromagnetic coupling between spins are good candidates for molecular magnets with highly stable magnetic moments under shape deformations. In particular, it was predicted that when these molecules have at least 20–50 transition metal ions (spin centers) arranged in 1–4 polygons, magnetic moments robust against molecular deformation can be achieved. Although such molecules are yet-to-be synthesized, it is hoped that this mathematical chemistry approach shown in this current work can be expanded and used to predict other molecular magnets and novel molecular materials.

Mathematical Methods and Tools Used in this Study:

Probability theory, stochastic stability and convergence, topology.

3.4.2 Mixing Time of Molecules Inside of Nanoporous Gold

*D.M. Packwood et al., SIAM J. Appl. Math. **74** (2014) 1298–1314.* [PJFCA]

3D network structures often appear in materials and greatly influence their properties and function. For example, electrical conductivity (migration of charge carriers) and the behavior of fluids (migration of dissolved components in fluid by diffusion/advection as well as the migration of fluid itself) are critically dependent upon the 3D network of conductive substances and pores in materials, respectively. In a study of 3D network structures, we always face the following three problems: first, acquiring and digitizing 3D structures, for which computed tomography, as explained in Chap. 2, is one of the typical tools used to obtain 3D image data (stacking data of many 2D slice images); second, extracting the characteristic features from 3D data sets; and third, extracting important parameters, from such data, which influence the properties of materials.

In this study, a team consisting of a theoretical chemist, a metallurgist, and synthetics chemists, investigated the 3D network structure of pores in nanoporous gold using graph theory and extracted the factors influencing diffusion in the nanopore network. Nanoporous gold is a two-component material consisting of metallic gold and a disordered network of pores. This material can be obtained by selective dissolution of silver from gold–silver alloy (this process is also called dealloying or selective leaching). The team used a 3D image of this material (Fig. 3.7a) obtained in their previous study using transmission electron microscope (TEM) tomography [Fuj]. They skeletonized pore parts in the 3D image using a thinning algorithm and extracted the graph of the nanoporous network (Fig. 3.7b). The graph shows only the connectivity of the porous network. Edges represent the pores and vertices represent where pores meet. Even though the graph is a highly simplified representation of an actual porous structure, it is still very complicated and difficult to describe. To describe the graph in a compact way, the team divided the graph into subgraphs and classified them. It turned out that the graph could be constructed from a large number of relatively simple types of small subgraphs. Some pores are isolated from the major connected pores and these isolated pores are removed from the data for diffusion analysis, which is explained below.

Using the simple subgraphs seen from the analysis above, they constructed a class of graphs that could be constructed from these subgraphs (called 'nanoporous graphs'), and studied the properties of random walks on these graphs. Some examples of these graphs are shown in Fig. 3.7c. Here, molecular diffusion is modeled by the random walks inside the pores of nanoporous gold. The 'speed' of diffusion

(a)   **(b)**   **(c)**

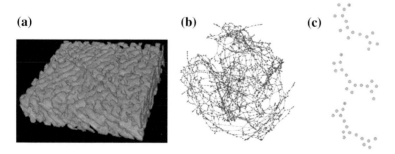

Fig. 3.7 a 3D image obtained by TEM-tomography. **b** 3D graph (pore network) extracted from the tomography data, **c** Three graphs of pore network used for numerical simulations. Reproduced from [PJFCA]. Copyright ©2014 Society for Industrial and Applied Mathematics. Reprinted with permission. All rights reserved

can be measured by the mixing time, which measures how quickly the probability distribution of the random walk converges to its equilibrium distribution, specifically

$$T_\varepsilon = \min\left(t : \max_{n \in G} |P(X_t = n) - P(X_\infty = n)| < \varepsilon\right), \tag{3.7}$$

where X_t is the position of the random walk at time t, and $P(X_t = n)$ is the probability that the random walk is located at vertex n at time t. Although an exact formula for the mixing time T_ε cannot be computed for the general case, the research team was able to calculate a good estimate of T_ε for the nanoporous gold graphs. This estimate showed that the mixing time could be reduced if a large number of small ring-like structures were included in the porous network [such as in the bottom two graphs in Fig. 3.7c], therefore, providing a novel insight into how 'local' connectivity of pores inside of nanoporous gold affects the rate of diffusion of molecules through the global pore network.

Mathematical Methods and Tools Used in this Study:

Graph theory
Probability theory and random walks (convergence rate of Markov chains)

References

[DS]   M. Deza, M. Dutour Sikirić, *Geometry of Chemical Graphs*, Encyclopedia of Mathematics and its Applications (Cambridge University Press, Cambridge, 2008)

[Fuj]   T. Fujita, L.H. Qian, K. Inoke, J. Erlebacher, M.W. Chen, Three-dimensional morphology of nanoporous gold. Appl. Phys. Lett. **92**, 251902–251905 (2008)

[Hir]   A. Hirata, L.J. Kang, T. Fujita, B. Klumov, K. Matsue, M. Kotani, A.R. Yavari, M.W. Chen, Geometric frustration of icosahedron in metallic glasses. Science **341**, 376–379 (2013)

[HNYI]   S. Hitosugi, W. Nakanishi, T. Yamasaki, H. Isobe, Bottom-up synthesis of finite models of helical (n, m)-single-wall carbon nanotubes. Nature Commun. **2** (2011). Article number: 492

[Iij]   S. Iijima, Helical microtubules of graphitic carbon. Nature **354**, 56–58 (1991)

[KHOCS]   H.W. Kroto, J.R. Heath, S.C. Obrien, R.F. Curl, R.E. Smalley, C60, Buckminster-fullerene. Nature **318** (1985) 162–163

[LPXY]   D.V. Louzguine-Luzgin, D.M. Packwood, G. Xie, AYu. Churyumov, On deformation behavior of a Ni-based bulk metallic glass produced by flux treatment. J. alloys compd. **561**, 241–246 (2013)

[MNHSKI]   T. Matsuno, H. Naito, S. Hitosugi, S. Sato, M. Kotani, H. Isobe, Geometric measures of finite carbon nanotube molecules: a proposal for length index and filling indexes. Pure Appl. Chem. **86**, 489–495 (2014)

[MT]   A.L. Mackay, H. Terrones, Diamond from graphite. Nature **352**, 762 (1991)

[NHHEMN]   T. Nakamura, Y. Hiraoka, A. Hirata, E.G. Escolar, K. Matsue, Y. Nishiura, Description of medium-range order in amorphous structures by persistent homology arxiv:1501.03611

[PJFCA]   D.M. Packwood, T. Jin, T. Fujita, M.W. Chen, N. Asao, Mixing time of molecules inside of nanoporous gold. SIAM J. Appl. Math. **74**, 1298–1314 (2014)

[PRFKT]   D.M. Packwood, K.T. Reaves, F.L. Federici, H.G. Katzgraber, W. Teizer, Two-dimensional molecular magnets with weak topological invariant magnetic moments: Mathematical prediction of targets for chemical synthesis. Proc. R. Soc. A **469**, 20130373 (2013)

[PSH]   D.M. Packwood, S. Shiraki, T. Hitosugi, Effects of atomic collisions on the stoichiometry of thin films prepared by pulsed laser deposition. Phys. Rev. Lett. **111**, 036101 (2013)

[SFDD]   R. Saito, M. Fujita, G. Dresselhaus, M.S. Dresselhaus, Electronic structure of chiral graphene tubules. Appl. Phys. Lett. **60**, 2204–2206 (1992)

[TLNKK]   M. Tagami, Y. Liang, H. Naito, Y. Kawazoe, M. Kotani, Negatively curved cubic carbon crystal with octahedral symmetry. Carbon **76**, 266–274 (2014)

Chapter 4
Breakthroughs Based on the Mathematics–Materials Science Collaboration

Abstract In this chapter, based on our experience at AIMR, Tohoku University, we describe the possible innovation that can be derived from mathematics–materials science collaborations. We believe that such collaborations need time to blossom, and once fields of common interest are integrated under a single interdisciplinary fusion, research will expand shared knowledge and advance science in new directions.

Keywords Interdisciplinary integration · New scientific fields · Hidden doors

4.1 Real Interdisciplinary Integration

As stated in previous chapters, in particular Chap. 3, we attempted to create a new materials science through a mathematics–materials science collaboration at our institute, AIMR, and some results have begun to emerge. These results contain novel points, which would have never been obtained without our interaction with mathematics. We are sure that such interdisciplinary collaboration has great potential to reveal unknown phenomena and open new scientific fields.

Through experience, we develop conditions that are necessary to derive real interdisciplinary fusion (integration). Here we suggest a mechanism of realization of the interdisciplinary integration using an analogy based on molecular orbital theory with orbital splitting that uses interorbital interaction. When one atom approaches another, two events are possible. One is that no interaction between the two atoms occurs, no chemical bond forms, and hence no molecule (Fig. 4.1a) forms. The other is that the atoms form a molecule. In this case, orbitals from the individual atoms interact with each other and produce new bonding and antibonding orbitals (Fig. 4.1b). Electrons tend to enter the bonding orbital, which has a lower kinetic energy (higher binding energy) to reduce the total energy of the system and therefore the two atoms bond as a molecule. We take the mechanism of formation of the new orbitals as a paradigm in the creation of new scientific disciplines through interdisciplinary research. The important point of this model is that such interactions create new orbitals (emerging disciplines) at different energy levels than the existing orbitals (established disciplines). Whereas the emerging discipline has a nature that is unclear before interaction, once

© The Author(s) 2015
S. Ikeda and M. Kotani, *A New Direction in Mathematics for Materials Science*, SpringerBriefs in the Mathematics of Materials,
DOI 10.1007/978-4-431-55864-4_4

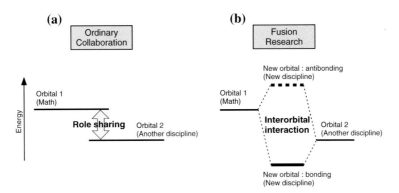

Fig. 4.1 Creation of "unknown" new scientific fields using an analogy based on molecular orbital theory of orbital splitting in an interorbital interaction

Fig. 4.2 Schematic showing a chemical reaction pathway

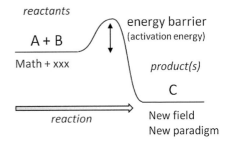

it has completely taken place, this discipline is sure to be a completely new scientific field that nobody has predicted.

To date, much joint research has been performed around the world. However, in most joint research, the style of collaboration is "role sharing" (Fig. 4.1a). For example, mathematicians take charge of the mathematical aspects and their collaborators take charge of their specialist areas. These collaborations may produce new results, but they do not create a new discipline or field based on such ordinary collaboration. What should be done to proceed to a higher level of integration? Researchers sometimes have to surmount barriers that divide disciplines as well as learn and assimilate fundamentals and special topics of the different disciplines. This task that is nevertheless enriching. With effort, some collaborations will accomplish real fusion and spur new research fields (Fig. 4.1b).

This process of integration can also be represented schematically as a chemical reaction $A + B = C$ (Fig. 4.2), where a brand-new chemical compound is produced. This new compound C is not a simple mixture of A and B but shows different characteristics from either. We can say that C corresponds to a new field or new paradigm created by interdisciplinary integration of A with B. While sounding somewhat simplistic, we should remember however that there is an energy barrier to overcome to produce the new compound. Clearly, we need to surmount the barriers that exist between the disciplines.

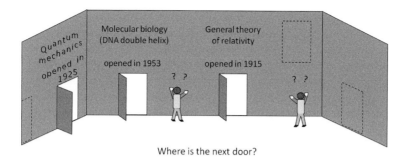

Where is the next door?

Fig. 4.3 We believe that hidden doors toward new scientific fields remain to be opened. Finding them is difficult. One possible path is to carry out interdisciplinary research that integrates fields much in the manner of the coupling of orbitals (Fig. 4.1b)

Figure 4.3 illustrates circumstances surrounding those struggling to find new scientific fields. Some doors have already been opened by great scientists. We believe some hidden doors toward new scientific fields still remain unopened. We want to find these doors but finding them requires strokes of genius. We believe that one possible way forward is to delve into interdisciplinary research. If integration is really achieved, we may then encounter those unknown orbitals (Fig. 4.1b). Because these orbitals have not been uncovered previously by researchers, their emergence in essence appears as new scientific worlds. In molecular orbital theory, the emerging orbitals produced by the interorbital interaction can be predicted by calculation. However, the analogy fails to a degree as it is difficult to predict where new scientific fields can or will exist. Therefore, we must encourage the exploration of interdisciplinary research regardless the risk of failure.

We mention two examples where scientific events became a turning point in science history based on the collaboration with mathematics. The first is the precedent set by study of relativity by Albert Einstein, one of the greatest physicists in human history. He noticed that "the bending of space" was the key to the theory of gravity and later was informed that Riemannian geometry could be applied to his theory. He studied and adopted Riemannian geometry to his new theory with the help of mathematician Marcel Grossmann, and finally completed the general theory of relativity. It should also be noted that David Hilbert, "father of modern mathematics", contributed to the completion of the general theory of relativity. After comprehending Einstein's work, Hilbert derived the equation of gravitational fields independently of Einstein (but keeping in touch with Einstein by mail). The competition with Hilbert accelerated Einstein's work. If we apply Fig. 4.1b to Einstein's accomplishment, it can be said that the special theory of relativity was discovered as a bonding orbital in 1905 by the interdisciplinary integration of electromagnetism and other topics of physics and mathematics. An excitation to antibonding orbital (discovery of the general theory of relativity in 1915) completed the integration with mathematics (Riemannian geometry and Hilbert's stimulation).

The second example is the precedent set by the research of Werner K. Heisenberg, who developed the basics of quantum mechanics. When he developed his theory, he hatched his calculation technique of using matrices. He professed not knowing matrix calculations but developed his technique to complete his new theory of quantum mechanics. This story provides a good lesson that we sometimes need to develop mathematical tools by ourselves to advance the forefront of science one step at a time when creating a new scientific field. Delta function and spin matrices created by Paul A.M. Dirac are also similar instances.

4.2 Organization Promoting Mathematics–Materials Science Collaboration

In the previous section, we mentioned our fundamental ideas to realize interdisciplinary integration, in particular, mathematics and materials science collaborations, and create new scientific fields. However, the scope of application of this idea is rather limited to collaborations between small groups because deep interaction seems difficult in large organizations. To spread the fruits of mathematics–materials science collaborations around the world, it is thus necessary to consider ways to promote mathematics–materials science collaboration on a bigger scale.

In this sense, we believe that our experience in mathematics–materials science collaboration in our institute, AIMR, is useful. Our institute successfully created a center where researchers from different backgrounds, including pure and applied mathematics, and languages can interact and communicate with each other. Referring to Fig. 4.4, exchanges among different fields begin from *interest*, and move toward phases of *interaction* and *inspire*, and finally reach *integration*. Through seminars, meetings, and collaborative research, we have climbed ladders, rung by rung, and achieved new vistas for mathematics–materials science collaboration. The key to accomplishing the mathematics–materials science collaboration is to find themes challenging for both materials scientists and mathematicians. We have identified several important challenges that have attracted and motivated both materials scientists and mathematicians. These include a comprehensive understanding of glassy materials, guiding principles for synthesis of carbon networks, universal understanding of the topological phase in spintronics, and fluidity of super highly concentrated particulate systems.

Generally, people tend to become hesitant because of great risks when setting out on new challenges with results that are difficult to obtain. However, the researchers in our institute are able to boldly tackle integration regardless of the risk of failure. This condition largely results from the researchers' spontaneous actions to achieve real integration, and the voluntary discussions and changes in attitudes prompted from such a cooperative environment.

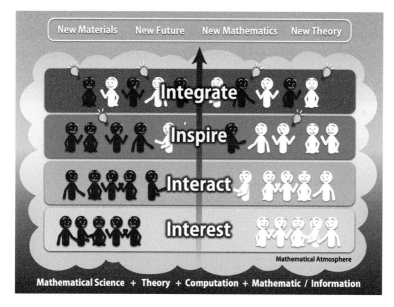

Fig. 4.4 Four steps of exchange toward integration. Courtesy of Ms. Miho Iwabuchi

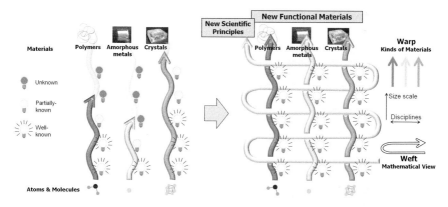

Fig. 4.5 The role of mathematics as wefts in textiles. The left/right figure shows the state before/after the threading of wefts (mathematics) through the warp (materials science). The authors thank Prof. Tadafumi Adschiri, Tohoku University, for providing us the idea of this figure

There is also an advantage in pushing mathematics–materials science collaboration at an institutional level. In a materials research institution, there are many researchers, each investigating different kinds of materials. This means that a lot of information and knowledge of various kinds of materials from the atomic/molecular level to the bulk material level is accumulated in one place. In the past, direct interaction in such a vast materials science field was difficult because the terminology is different in each field and we do not have time to review and process results for all these materials. In contrast, mathematics can play a role as wefts (the yarn woven across the warp yarn in weaving) in a textile with materials as the warp (the woof; vertically aligned yarn) of a textile (Fig. 4.5). For each warp, understanding the structure and property of a material has gradually progressed over the centuries. However, the understanding of some layers in the hierarchy remains missing. Mathematics as wefts will connect these layers by providing a common language for all fields and help to find common structures and properties at each level of the hierarchy. All such efforts will lead us to a comprehensive understanding of materials systems and the creation of a predictive materials science in the future.

4.3 Specific Problems and Applications in the Future

In the *Odom Report* described in Chap. 2, several areas were identified where mathematics in society might have an impact. These are listed in the following:

Problem/Application	Contribution from Mathematics
MRI and CAT	Imaging integral geometry
Air traffic control	Control theory
Options valuation	Black–Scholes options model and Monte Carlo simulation
Global reconnaissance	Signal processing, image processing, data mining
Stockpile stewardship	Operations research, optimization theory
Stability of complex networks	Logic, computer science, combinatorics
Confidentiality and integrity	Number theory, cryptology/combinatorics
Modeling of atmospheres and oceans	Wavelets, statistics, numerical analysis
Agile, automated manufacturing	Geometry, visualization, robotics, control theory, in process quality control
Design and training	Simulation, modeling, discrete mathematics
Analysis of the human genome	Data mining, pattern recognition, algorithms
Rational drug design	Data mining, combinatorics, statistics
Seiberg–Witten questions (string theory)	Geometry
Interpreting data on the universe	Data mining, modeling, singularity theory
Design systems for composite materials	Control theory, computation, partial differential equations
Earthquake analysis and prediction	Statistics, dynamical systems/turbulence, modeling, in process control

In materials science, designing new materials based on mathematical theory, for example, control theory, computation, and partial differential equations, is suggested. This is still an important challenge and will continue to be the main emphasis in mathematics–materials science collaborations. Furthermore, we are also facing a problem that we have to analyze Big Data and extract important information from huge databases to realize a predictive materials science. Considering this aspect, the importance of mathematics is evident. However, contrary to such comprehensive trends, each concrete problem in materials science is complicated and involves

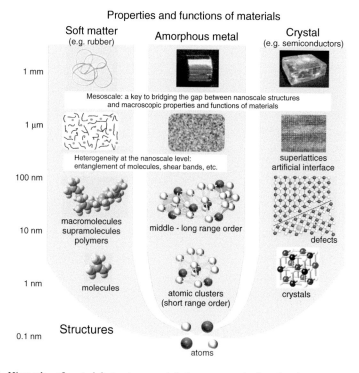

Properties and functions of materials

Soft matter (e.g. rubber)   Amorphous metal   Crystal (e.g. semiconductors)

1 mm

Mesoscale: a key to bridging the gap between nanoscale structures and macroscopic properties and functions of materials

1 μm

Heterogeneity at the nanoscale level: entanglement of molecules, shear bands, etc.   superlattices artificial interface

100 nm

macromolecules supramolecules polymers   middle - long range order

10 nm

defects

1 nm

molecules   atomic clusters (short range order)   crystals

0.1 nm   **Structures**

atoms

Fig. 4.6 Hierarchy of material structures and their macroscopic functional expression based on microscopic structures. The cross-sectional image of a superlattice was provided by courtesy of Prof. Masashi Kawasaki, The University of Tokyo

a broad spectrum of topics. Figure 4.6 shows a hierarchy of structure–property relationships from the atomic/molecular level to the bulk material level. Our ultimate goal is to understand comprehensively all relationships between individual hierarchical layers and the types of materials and predictively design microscopic structures which express the desired macroscopic properties and functions by solving inverse problem. However, there are a lot of discontinuities or non-linear connections between them and we have not succeeded in bringing all layers together. First of all, we have to consider in depth each relationship between the layers through thorough discussions between mathematicians and materials scientists and gradually connect the layers and materials to accomplish a comprehensive understanding of materials and a predictive materials science. Unexplored challenges are always accompanied by risk. However, they are also accompanied by exuberance and anticipation. As suggested in this book, the next breakthrough is expected to be achieved based on interdisciplinary integration and, in particular, a mathematics–materials science collaboration hopefully will lead this trend and create new scientific fields based on predictive design and development of materials that will contribute to society in the not too distant.

Chapter 5
Epilogue

By reviewing relevant historical events and facts as well as some results achieved at AIMR, we argued in this book for the importance of a mathematics–materials science collaboration as progenitor to the next breakthroughs and innovations. However, we suppose that not all readers of this book will agree to our argument because materials science has always been an empirical field where theory tends to follow experimental results. The establishment of a "predictive" materials science might seem difficult to achieve. Nevertheless, we offer one further historical story where a theoretical prediction preceded experiments.

The story relates the discovery of screw dislocation and spiral growth of crystal surfaces by Charles Frank [BCF]. The fundamental theory of crystal growth had been almost completed prior to 1950. However, the theory could not explain the large growth rate in actual systems under certain supersaturation conditions. Frank noticed that the actual growth rate could be explained if screw dislocations and spiral growth were possible. Frank asked W.K. Burton and N. Cabrera to calculate the growth rate with the new growth model and obtained the expected result [BCF]. After Burton, Cabrera and Frank presented their theory at the Faraday Discussion on Crystal Growth in Bristol in 1949, for a while, the response from researchers was complicated because nobody had seen such spirals on crystal surfaces. However, L.J. Griffin, who heard Frankfs new theory, succeeded in obtaining beautiful growth spirals on the crystal surfaces of beryl using the phase-contrast microscope. Pictures were shown at a subsequent conference in Bristol, where they aroused great excitement. After Griffin's work, many experimental results of spiral growth were reported from around the world [Cah].

In materials science, experiments tend to precede theory. However, in the above story, theory indicated an unexpected model and opened a new field of crystal growth. We believe that this is one example supporting our proposal. Although intended readership of this book is mainly mathematicians, we hope that this book will also stimulate materials scientists to join the network of mathematics–materials science collaborations and create new materials science together.

© The Author(s) 2015
S. Ikeda and M. Kotani, *A New Direction in Mathematics for Materials Science*, SpringerBriefs in the Mathematics of Materials, DOI 10.1007/978-4-431-55864-4_5

References

[BCF] W.K. Burton, N. Cabrera, F.C. Frank, The growth of crystals and the equilibrium structure of their surfaces. Phil. Trans. Roy. Soc. Lond. **234A**, 299–358 (1951)

[Cah] R.W. Cahn, *The Coming of Materials Science*, Pergamon materials series (Pergamon, Amsterdam, 2001)

Appendix A
Supplements to "Quantum Materials"

Here we cover basic matters of semiconductors and spintronics that were not explained in the main text.

A.1 Semiconductors

The greatest invention in electronics in the 20th century is the "transistor". Fabricated at Bell Laboratories in 1947, a transistor is an electronic device that can amplify or switch on/off electrical power. A semiconductor is used as the main component of a transistor. Before the invention of a transistor, vacuum tubes had been used to amplify, rectify, oscillate or switch on/off electrical power. These functions are indispensable in constructing electronic device circuits. Metals can carry electric currents but they cannot stop them. In contrast, insulators cannot carry currents, but can block them. We need a material that can both carry and block currents if we want to realize electronic device circuits using "materials" (without vacuum tubes). Even before the invention of a transistor, it had been known that there exist three types of material: conductors (metals), semiconductors, and insulators. Their definitions are based on the electrical conductivity (or resistivity) of the materials. The boundary between conductors and semiconductors is roughly defined as 10^5 S/m (10^{-5} Ωm), whereas the boundary between semiconductors and insulators is roughly defined as 10^{-4} S/m (10^4 Ωm). Before 1947, however, researchers other than those at Bell Laboratories did not believe that semiconductors would be used usefully in devices to control electrical power. John Bardeen, Walter Brattain, and William Shockley at Bell Laboratories succeeded in creating a point-contact germanium device, the famous first transistor in the world. After their invention, various transistor-based products such as hand-held radios followed, which had an impact on daily life. With the size and price of electronic devices becoming ever smaller, computers were brought into the home and personalized. More recently, much of a computerfs functioning has been compacted into hand-held devices such as smart phones.

© The Author(s) 2015
S. Ikeda and M. Kotani, *A New Direction in Mathematics for Materials Science*, SpringerBriefs in the Mathematics of Materials, DOI 10.1007/978-4-431-55864-4

The mechanism to control current using a semiconductor is a kind of magic. A semiconductor by itself is not capable of controlling current; its electrical conductivity is lower than that of metals but larger than that of insulators. The function of controlling current is the result of cooperation with device technology (e.g., [Kitt, Sze06, Sze01]). Below, we recapitulate the basic mechanism of switching current with semiconductors. For pure (called "intrinsic") silicon (Si), the typical semiconductor for electronics, the electrical conductivity (or resistivity) is around 10^{-3} S/m (10^3 Ωm) at room temperature, which is close to the value of insulators and not enough to carry current. Therefore, we add impurities ("dopants") such as phosphorus (P; having a valence of five) or boron (B; having a valence of three) to silicon (Si; having a valence of four) to produce electrons (negative charges) or holes (positive charges), called "charge carriers," so that the electrical conductivity increases. The impurities are called "donor (which donates an electron)" and "acceptor (which accepts an electron and leaves a hole)"; semiconductors with donor/acceptor atoms are called "n-type"/"p-type". In n-type semiconductors, electrons are majority carriers while holes (positive charges) are the minority carriers. In p-type semiconductors, the reverse occurs where holes are the majority carriers while electrons are the minority carriers. Once we succeed in increasing the electrical conductivity, we next have to consider how we block the electrical current to achieve the on/off switching that is necessary to realize electronic circuits. Here, we need to create a device structure for this purpose. A typical structure (Fig. A.1) of a field effect transistor (FET) uses

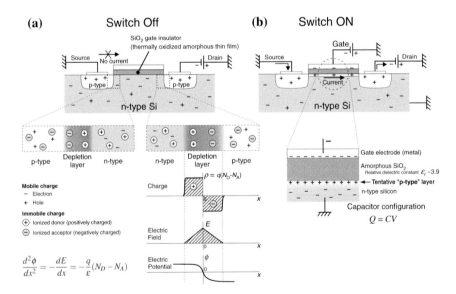

Fig. A.1 Cross-sectional illustration of a p-channel silicon FET fabricated using an n-type silicon wafer: **a** for $V_G = 0V$ and **b** for $V_G < 0V$. Holes gather along the semiconductor (n-type silicon)/gate insulator (SiO_2) interface by the field effect (just like a capacitor) and a hole conduction channel (source-channel-drain) is formed

an n-type silicon wafer (charge carriers are electrons from doping with such as P). For the p-type regions (charge carriers are holes) by implantation with such as boron. These doped regions can be made, for example, using the ion beam technique. We also have to make an amorphous SiO_2 thin film as a gate dielectric layer on the surface of the substrate (for example, by thermal oxidation of silicon) and attach metal electrodes for source, drain and gate. First we consider applying just a voltage to the drain electrode and no voltage to the gate electrode (Fig. A.1a). Inside the silicon substrate, there are two boundaries between the n-type and p-type regions; such a structure is called a "p-n junction". In this junction, a fraction of the electrons in the n-type region diffuses into the p-type region and a fraction of the holes diffuses into the n-type region, and these carriers disappear by recombination. Subsequently, a thermal equilibrium state is achieved and a so-called "depletion layer (region)," meaning the absence of charge carriers in the layer. The donors that have lost electrons are positive ions at the top of the n-type region. Similarly, the acceptors that have lost holes are negative ions at the top of the p-type region. An electric field therefore appears in the depletion layer. This state can be described by Poisson's equation:

$$\nabla^2 \phi = -\nabla \mathbf{E} = -\frac{\rho}{\varepsilon}, \tag{A.1}$$

where ϕ is the electrical potential, E the electric field, ρ the charge density, and ε the dielectric constant (ε is the product of the dielectric constant under vacuum ε_0 and the relative dielectric constant of the material ε_r). In the device, as the current is a one-dimensional flow of charge, we can reduce the dimension from 3D to 1D, and the above Poisson's equation can be expressed as

$$\frac{d^2\phi(x)}{dx^2} = -\frac{dE(x)}{dx} = -\frac{q}{\varepsilon}(N_D - N_A + p - n), \tag{A.2}$$

where q is the charge of an electron, N_D the concentration of ionized donors, N_A the concentration of ionized acceptors, p the concentration of holes, and n the concentration of electrons.

In the depletion layer, the p and n-type regions have almost zero charge as they disappear by recombination. Hence, Eq. (A.2) can be simplified as

$$\frac{d^2\phi(x)}{dx^2} = -\frac{dE(x)}{dx} = -\frac{q}{\varepsilon}(N_D - N_A). \tag{A.3}$$

Figure A.1a shows the condition where no voltage is applied to the gate electrode ($V_G = 0V$). In this case, there are two p-n junctions and the electric field generated in the depletion layers at the junctions block the current, i.e., the switch off state. Strictly speaking, there is some difference in the thicknesses of the depletion layers and their capability to block current for the p-n junction because of the direction of the bias application. In contrast, similar to the effect of a capacitor, holes accumulate along the gate insulator (SiO_2)/semiconductor (Si) interface and this region resembles a p-type silicon (Fig. A.1b with $V_G < 0V$). Once the three p-type regions are connected, the connected path works as a conduction channel (i.e., the switch on state).

As we see in the above discussion, functionality of semiconductor devices is produced through the combination of electronic properties of the semiconductor materials and the "geometrical" configuration of devices. Clearly, geometrical aspects need to be taken into account.

Anderson Localization

Next, in connection with semiconductors, we consider the problem of impurity doping. We dope semiconductors with impurities to produce carriers and increase conductivity. However, if the concentration of impurities increases, the scattering probability of carriers by the impurity atoms increases, and this affects the electric conductivity. Therefore, doping should be optimized keeping this balance at an appropriate level. With the increase in impurities, the mechanism of electron transport departs from that underpinning **Bloch's theorem** ((1.2) and (1.3)) described in the section "Solid State Physics" in Chap. 1. This problem reduces to **Anderson localization** proposed by Philip W. Anderson, 1977 Physics Nobel laureate [And].

A square-well potential is a typical problem in quantum mechanics. If an "infinitely deep" square-well of width L exists (Fig. A.2a), a particle is completely confined in the well and the solution of Schrödinger equation takes the well-known form

$$\psi_n(x) = \sqrt{\frac{2}{L}} \sin k_n x = \sqrt{\frac{2}{L}} \sin \frac{n\pi x}{L} \quad \text{where} \quad n = \{1, 2, 3, 4, \ldots\}. \quad \text{(A.4)}$$

For a square-well of finite depth (Fig. A.2b), the particle probability is everywhere finite and the wave function far from the well can be expressed by the following two forms [Kaw]:

$$\psi_s(x) = Ae^{ikx} + Be^{-ikx} \quad \text{(scattering state)} \quad \text{(A.5)}$$

$$\psi_b(x) = Ce^{-ik|x|} \quad \text{(bound state).} \quad \text{(A.6)}$$

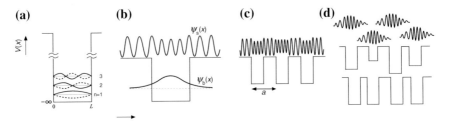

Fig. A.2 Wave functions of particles for various types of square-well potentials: **a** infinite depth, **b** finite depth, **c** a line of wells with the same size and with exact periodicity a, and **d** a line of wells with different depths or the same size of wells but without exact periodicity

If square wells of the same width and the same depth form a one-dimensional line with exact periodicity of interval a (Fig. A.2c), eigenstates that are not attenuated by the potential distribution can exist and the wave function becomes

$$\phi_l(x) = \sum_{n=0}^{N-1} e^{2i\pi ln/N} \psi(x - na), \quad (l = 0, 1, \ldots, N - 1). \tag{A.7}$$

What happens, then, if these square wells are of different depths or are of the same size but without exact periodicity (Fig. A.2d)? Anderson proposed this problem and suggested the possibility that a "localized state" can coexist with an "extended state" in systems with disorder [And]. However, his suggestion was rather qualitative and researchers had to wait for **scaling theory of localization** published by E. Abrahams, P.W. Anderson, D.C. Licciardello, and T.V. Ramakrishnan in 1979 [AALR] for a further quantitative treatment of Anderson localization. We do not explain their theory here, but the basic scaling parameter in their theory has form

$$\beta(g(L)) \equiv \frac{d \log g(L)}{d \log L}, \tag{A.8}$$

where L and $g(L)$ are, respectively, the size and conductance (the reciprocal of resistance $[\Omega^{-1}]$) of the material. If the material is a perfect crystal without any defect, the wave function spreads out over the crystal without any break and the conductance should simply decrease inversely proportional to L (if the thickness of the material is constant). In other words, the resistivity increases linearly with L. However, if there are localized states, depending on the size (localization length) of the eigenstate, the conductance becomes scale-dependent. To date, many experiments have been performed following confirmation of Anderson localization before the proposal of the scaling theory.

A.2 Spintronics

"**Spintronics**" is a portmanteau that abbreviates and combines phrases "spin transport electronics" and "spin-based electronics." As stated earlier, electronic devices using semiconductors have changed daily life completely. In device technology, especially involving silicon-based materials, downsizing of devices has increased operating speed and decreased energy loss due to Joule heating by current flow. However, we are facing limitations in downsizing through the appearance of quantum effects (e.g., tunneling of charges), and we can no longer simply reduce component sizes. Furthermore, as long as we use "charge" as a medium for transferring information, energy loss because of Joule heating is inevitable. It is known that as well as "charge" an electron has another degree of freedom called "spin," and the idea immediately surfaces of using "spin" for transferring information instead of charge. Pure spin currents do not produce any Joule heat and devices with low energy dissipation can be realized. However, the relaxation length of a spin ranges from several tens of nanometers to sub-micrometer and it was technically difficult to use spin to transfer information. In the past one to two decades, micro-/nano-fabrication techniques have largely developed and short spin relaxation lengths have become within levels that can be handled. Spintronics nowadays covers a wide range of research fields, but here we briefly explain those fields that have been recognized as almost independent.

Giant Magnetoresistance

Magnetism has always been a major topic in physics (in solid state physics and geophysics) and materials science. Historically, the creation of the strongest magnets has been a major objective in the research field. This historical trend was divided into two branches, one being to further the advance of creating the strongest magnet (controlling collective phenomenon of spin) and the other being spin-based electronics (controlling individual spins) following the independent discovery in 1988 of "giant magnetoresistance (GMR)" by the group of Albert Fert [BBF] and the group of Peter Grünberg [BGSZ]. GMR is a magnetic phenomenon whereby the electric resistance of materials changes (up to 50 % or more) with the application of an external magnetic field. The materials are generally multilayer thin films composed of alternating layers of ferromagnetic/non-magnetic compounds. There exist two layers, a fixed layer (whose spin direction is fixed) and a free layer (whose spin direction is changeable by the external magnetic field). If the spin direction in the free layer is adjusted to that of the fixed layer by the external field, the electrical resistance of the device decreases. If the spin direction in the free layer opposes the external field, the electrical resistance increases. GMR was exploited in magnetic field sensors to read data recorded in hard disks. This application turned out to be a great success in downsizing hard disks and Fert and Grünberg received the Nobel Prize in Physics in 2007. It is widely recognized that the discovery of GMR was the starting point of "spintronics."

Tunnel Magnetoresistance

Tunnel magnetoresistance (TMR) sounds similar to GMR, but the effect and mechanism is different from those of GMR. The device structure that features the TMR effect is simpler than that of GMR devices; an insulator thin film is sandwiched between two ferromagnetic layers (i.e., two ferromagnetic layers are separated by an insulator thin film, which prevents short circuiting). If the spin direction of the two ferromagnetic layers is the same, the probability of electron tunneling through the insulating film increases and electrical resistance decreases. After the discovery of GMR, Terunobu Miyazaki focused on TMR, which was believed to occur only at very low temperatures, and in 1994 realized an 18 % TMR ratio at room temperature using a magnetic tunneling junction of Fe/Al_2O_3(amorphous)/Fe. In 1995, two papers reporting a TMR effect at room temperature were published independently, igniting TMR research around the world [MT, Moo]. Much effort has gone into applying TMR to the creation of non-volatile magnetoresistive random-access memory (MRAM), which will provide energy savings in computers.

Magnetic Semiconductors

At the end of the 1990s, a new material series "magnetic semiconductors" was created. Electronic circuits and devices are classified as those controlling charge current for transferring information and driving displays, this role now being played by semiconductors, and the other is to store and record information, the role being played by magnetic materials (spin). If we can control both charge and spin in one material, we can decrease the number of electronic components and integrate them in the same circuit, thus reducing the energy dissipation. Hideo Ohno tried to grow new III–V-based diluted magnetic semiconductors (DMSs) using MBE, succeeding with ferromagnetic (Ga,Mn)As in 1996 [Ohn98]. In 2000, Ohno et. al. succeeded in controlling magnetism within a magnetic semiconductor (In,Mn)As by changing the hole concentration with the field effect (using a FET device, Fig. A.1) [Ohn00]. A theoretical model explaining magnetic semiconductors was proposed by Dietl et al. also in 2000 [DOMCF]. Combinations of semiconductors and magnetic properties are a big challenge in realizing for example non-volatile semiconductor memories and high-speed electronic circuits.

Spin Current and Spin Caloritronics

The research field of "spin current" is developing quickly with many new ideas and discoveries. The **spin Hall effect (SHE)** was theoretically predicted by M.I. Dyakonov and V.I. Perel in 1971 [DP] and after gaining a deeper understanding of the theory [MNZ03], SHE was experimentally observed in semiconductors by D.D. Awschalom's group in 2004 [KMGA]. In SHE, spin current is produced by a charge current in non-magnetic metals and semiconductors. If we have an opposing phenomenon, that is, one in which current is produced by a spin current, it is useful for detecting spin currents and for energy harvesting. The **inverse spin Hall effect (ISHE)** was first observed in a $Ni_{81}Fe_{19}/Pt$ sample in 2006 by Eiji Saitoh's group [SUMG]. They also found that heat (temperature difference) produces a spin current and that we can obtain charge currents through ISHE. This process in which electricity is generated by heat through the ISHE is called the **spin Seebeck effect** and was first observed by Saitoh's group in 2008 [Uch]. Using the spin Seebeck effect, we can obtain electricity from a very simple structure using a thin film [Kir]. Such new trends using spin current for energy harvesting is nowadays called **spin caloritronics** [BSvW].

References

[AALR]   E. Abrahams, P.W. Anderson, D.C. Licciardello, T.V. Ramakrishnan, Scaling theory of localization: absence of quantum diffusion in two dimensions. Phys. Rev. Lett. **42**, 673–676 (1979)

[And]   P.W. Anderson, Absence of diffusion in certain random lattice. Phys. Rev. **109**, 1492–1505 (1958)

[BBF]   M.N. Baibich, J.M. Broto, A. Fert, F. Nguyen Van Dau, F. Petroff, P. Eitenne, G.
           Creuzet, A. Friederich, J. Chazelas, Giant magnetoresistance of (001)Fe/(001)Cr
           magnetic superlattices. Phys. Rev. Lett. **61**, 2472–2475 (1988)
[BSvW]   G.E.W. Bauer, E. Saitoh, B.J. van Wees, Spin caloritronics. Nature Mater. **11**, 391–
           399 (2012)
[BGSZ]   G. Binasch, P. Grünberg, F. Saurenbach, W. Zinn, Enhanced magnetoresistance in
           layered magnetic structures with antiferromagnetic interlayer exchange. Phys. Rev.
           B (rapid communications) **39**, 4828–4830 (1989)
[DOMCF]   T. Dietl, H. Ohno, F. Matsukura, J. Cibert, D. Ferrand, Zener model description of
           ferromagnetism in zinc-blende magnetic semiconductors. Science **287**, 1019–1022
           (2000)
[DP]   M.I. Dyakonov, V.I. Perel, Current-induced spin orientation of electrons in semicon-
           ductors. Phys. Lett. A **35**, 459–460 (1971)
[KMGA]   Y. Kato, R.C. Myers, A.C. Gossard, D.D. Awschalom, Observation of the spin Hall
           effect in semiconductors. Science **306**, 1910–1913 (2004)
[Kaw]   A. Kawabata, Anderson localization and mesoscopic systems (in Japanese), in ed.
           by H. Fukuyama, K. Yamada, and T. Ando. Solid State Physics for Graduate School
           Students, vol. 3 (Kodansha Scientific Books, 1996)
[Kir]   A. Kirihara, K. Uchida, Y. Kajiwara, M. Ishida, Y. Nakamura, T. Manako, E. Saitoh,
           S. Yorozu, Spin-current-driven thermoelectric coating. Nature Mater. **11**, 686–689
           (2012)
[Kitt]   C. Kittel, *Introduction to Solid State Physics*, 8th edn. (Wiley, New York, 2004)
[MT]   T. Miyazaki, N. Tezuka, Giant magnetic tunneling effect in Fe/Al$_2$O$_3$/Fe junction.
           J. Magn. Magn. Mater. **139**, 231–234 (1995)
[Moo]   J.S. Moodera, L.R. Kinder, T.M. Wong, R. Meservey, Large magnetoresistance at
           room temperature in ferromagnetic thin film tunnel junctions. Phys. Rev. Lett. **74**,
           3273–3276 (1995)
[MNZ03]   S. Murakami, N. Nagaosa, S.-C. Zhang, Dissipationless quantum spin current at
           room temperature. Science **301**, 1348–1351 (2003)
[Ohn98]   H. Ohno, Making nonmagnetic semiconductors ferromagnetic. Science **281**, 951–
           956 (1998)
[Ohn00]   H. Ohno, D. Chiba, F. Matsukura, T. Omiya, E. Abe, T. Dietl, Y. Ohno, K. Ohtani,
           Electric-field control of ferromagnetism. Nature **408**, 944–946 (2000)
[SUMG]   E. Saitoh, M. Ueda, H. Miyajima, G. Tatara, Conversion of spin current into charge
           current at room temperature, inverse spin-Hall effect. Appl. Phys. Lett. **88**, 182509
           (2006)
[Sze01]   S.M. Sze, *Semiconductor Devices: Physics and Technology*, 2nd edn. (Wiley, New
           York, 2001)
[Sze06]   S.M. Sze, K.Ng Kwok, Physics of Semiconductor Devices, 3rd edn. (Wiley-
           Interscience, New York, 2006)
[Uch]   K. Uchida, S. Takahashi, K. Harii, J. Ieda, W. Koshibae, K. Ando, S. Maekawa, E.
           Saitoh, Observation of the spin Seebeck effect. Nature **455**, 778–781 (2008)

Printed in the United States
By Bookmasters